Boron Neutron Capture Therapy

Toward Clinical Trials of
Glioma Treatment

Boron Neutron Capture Therapy

Toward Clinical Trials of
Glioma Treatment

Edited by

Detlef Gabel

University of Bremen
Bremen, Germany

and

Ray Moss

Joint Research Centre
Petten, The Netherlands

Editorial Assistant
Renate Alberts

SPRINGER SCIENCE+BUSINESS MEDIA, LLC

Library of Congress Cataloging in Publication Data

Boron Neutron CaptureTherapy: toward clinical trials of glioma treatment / edited by
Detlef Gabel and Ray Moss.
    p.   cm.
    "Proceedings of an international workshop and plenary meeting . . . held September
18–20, 1991, in Petten, the Netherlands"—T.p. verso.
    Includes bibliographical references and index.
    ISBN 978-1-4613-6506-8   ISBN 978-1-4615-3408-2 (eBook)
    DOI 10.1007/978-1-4615-3408-2
    1. Gliomas—Radiotherapy—Congresses. 2. Boron-neutron capture therapy—Con-
gresses. I. Gabel, Detlef. II. Moss, Ray.
    [DNLM: 1. Boron Compounds—pharmacology—congresses. 2. Boron Compounds
—therapeutic use congresses. 3. Glioma—radiotherapy—congresses. 4. Neutrons—
therapeutic use—congresses. 5. Radiotherapy, High-Energy—methods—congresses.
QZ 380 B737 1991]
RC280.B7B68   1992
616.99′4810642—dc20
DNLM/DLC   92-48775
for Library of Congress   CIP

Proceedings of an international workshop and plenary meeting entitled
Towards Clinical Trials of Glioma with Boron Neutron Capture Therapy,
held September 18–20, 1991, in Petten, The Netherlands

ISBN 978-1-4613-6506-8

© 1992 Springer Science+Business Media New York
Originally published by Plenum Press, New York in 1992
Softcover reprint of the hardcover 1st edition 1992

PREFACE

The European Collaboration on Boron Neutron Capture Therapy (BNCT), conceived in 1987 and successful in 1989 in gaining financial support as a Concerted Action through the Medical and Health Research Programme of the Commission of the European Communities (CEC) in Brussels, considered it an opportune moment to hold its annual Plenary Meeting on 18-20 September 1991 as an International Workshop entitled "Towards Clinical Trials of Glioma with BNCT".

The background to this consideration was influenced by the world-wide resurgence of interest in NCT over the last 2 decades and by the exemplifications at the Fourth International Symposium on Neutron Capture Therapy for Cancer held in Sydney in December 1990, where it was strongly indicated that within the next 2 years clinical trials would be started both in Europe and the United States. In particular at the High Flux Reactor of the Joint Research Centre of the CEC at Petten in The Netherlands, an epithermal neutron beam designed and installed in the summer of 1990 recently became operable at full reactor power. An extensive series of experiments, including the nuclear and radiobiological characterisation of the beam and a healthy tissue tolerance study on canines has started and has the aim to define the preconditions for clinical trials on patients with Grade III/IV glioma.

However, as with any other new therapy modality, it must be demonstrated that BNCT is safe for the patient and has a reasonable chance of being an effective therapy. Consequently, it was felt appropriate that this International Workshop is held to review all available data for treatment and derive a consensus for a strategy for clinical trials which can meet with the agreement of the scientific and medical community. Without such support, the future of BNCT world-wide could be seriously jeopardised. The Workshop highlights also some perspectives of future developments of BNCT and on the background of clinical trials.

At this International Workshop, over 100 experts world-wide, including invited speakers from the United States, Australia and Japan, as well as participants from most European Countries, met to discuss BNCT at the location of the Institute for Advanced Materials of the Joint Research Centre of the European Communities at Petten.

Finally, we would like to thank Dr. A. Vermorken and the CEC for the continuing support of the European Collaboration for Boron Neutron Capture Therapy and of this Workshop, and the staff of the Joint Research Centre Petten and the ECN Petten for help with the organisation.

Detlef Gabel   Ray Moss

23 October 1992

CONTENTS

Contents

OPENING SPEECH

E.D. Hondros

Commission of the European Communities
Institute for Advanced Materials
Joint Research Centre
NL-1755 ZG Petten, The Netherlands

Good morning and a warm welcome to you all - participants from all parts of the European Communities, of Europe and other countries including Japan, USA and Australia.

The subject of our meeting today is a rather rare one for us here, Boron Neutron Capture Therapy. It is a fascinating subject and in many ways, it is truly interdisciplinary in that it joins together specialists from many professions, from neutron physicists on the one hand and to neurosurgeons on the other, and including chemists, radiobiologists in-between. Therefore, a truly multidisciplinary and a very challenging subject. It is exciting, not only to the specialists, but to everybody one talks to. This is a rare subject matter, in that it can lead to benefit which would be felt by people directly.

How is it that our centre, which is really the Institute for Advanced Materials of the Joint Research Centre of the European Communities, in a bizarre manner embraces activities such as boron neutron capture therapy when our normal métier is engineering materials? Together with our neighbours at the Dutch institute of ECN, we welcome this activity because it is a way of optimising the application of neutrons derived from our fine machine, the High Flux Reactor, into areas which are socially acceptable. This is a very rare experience these days, because nuclear things appears to be, for reasons we need not go into and usually for the wrong reasons, received in an unfriendly way. BNCT is a nuclear friendly area. We therefore welcome the opportunity to encourage this application of the HFR in such fields.

I would like to say a few words about the European Communities and its role in research. You may know that the Commission of the European Communities sponsors research through its Framework Programme. Among the main mission lines of the programme, there is the one which relates to improving the health and safety of the citizens of Europe and here funds are made available for doing research in fields such as cancer, AIDS and so on. It

Boron Neutron Capture Therapy, Edited by D. Gable
and R. Moss, Plenum Press, New York, 1992

is this particular segment of the Framework Programme which is supporting the Concerted Action on BNCT, which is represented by many of you here today. I need not go into details on how the rest of our Institute is being sponsored by the European Communities, but it is little related with the medical matters.

How is it that we have developed an interest in this field? Historically in Petten, the nuclear interests have been connected with the effects of irradiation on engineering materials used in the nuclear industry. Our staff for years have been in a peripheral way thinking about this subject of BNCT, which, as you know very well, has been around for almost 50 years. Our staff have often talked about it. It was in about 1987, when following enquiries and contacts, it was found that here in Petten we have one of the few research reactors that is uniquely suitable, with the right neutron beam capabilities, for studying BNCT. This fitted in very well with our policies and we, of course, reacted very strongly to this possibility and tried to find support, which has been very difficult, but fortunately through our Exploratory Research Programme of the Joint Research Centre, we have been able to find support for a few members of staff and specific credits to keep the subject alive. We tried every avenue and have been helped also by the same Directorate which is supporting you in the Concerted Action.

So far, direct contributions from the European Communities have been minor. But, let me remind you, this is not an indicator of the extent of interest in the subject. Over the past 3 years, a neutron beam has been designed, manufactured and installed, and I have just been informed is being operated at full reactor power since July of this year. Therefore, the HFR in Petten our colleagues at ECN are in a strong position to continue to make a contribution to this BNCT research. Our primary goal will be to undertake clinical trials within the next couple of years. The outcome of the clinical trials will, I believe, be critical to the whole future of BNCT because if these are proved to be highly promising, I am convinced, as much as one can be on these matters, that funds will be made available.

I have to emphasise the existence of the Concerted Action which is a unique means for conducting research. I like the Concerted Action format, because people join because they believe in it. The money that is being made available from Brussels is just "lubrication" funds, minuscule amounts for helping to hold meetings and support the travel. I hope that in the future this form of budgeting will be expanded into a true budget for research and clinical operations. I note also the presence of a strong international cooperation. This is a great future, which is reflected in the origins of some of our guests present here and, of course, the similar meetings, similar workshops that have been conducted in other parts of the world.

Finally, I come back and refer to the bizarre element in all this - why should this essentially medical matter be hosted in our Centre, which is an engineering and materials centre? I have mentioned the touch of opportunism about it, the fact the HFR is uniquely capable of being employed in this sort of study and we have responded to it. But I have noted another element in all this, mainly our staff. Our staff are truly motivated by a type of spirit which one does not see often among engineers. I can quite rightly call it a humanitarian spirit. They believe that engineers also can contribute something that can help mankind directly. The benefits of their research is not a long winded procedure - it is a direct benefit. That seems to charge emotionally our staff. I believe strongly that they are very sincere in their feelings, in their motivation to do this work. Not only is it a technical and scientific challenge, but we notice that it may have some immediate humanitarian benefit, which thrills us because it is a rare experience in the engineering sciences.

Finally, our responses within our institute are quite positive for reasons I have indicated. But even more important, the responses among the people in the streets who are occasionally fed with this sort of publicity are extremely positive. That is, people in the Netherlands, people in the locality, people in the local schools who ask us what are we up to - we can say that we pursue ideas experimentally and some ideas are such and such, and let me tell you, this is one idea which excites them.

But above all, it excites the politicians. That is a hard thing to do these days, and it is a very important thing to be able to excite the politicians because they are the ones who control the budgets. In discussions in Brussels, I am often asked how is that BNCT business going on in Petten? Are we going to see any results in the future? Thereby hangs the important aspect of credibility. It is a political problem to convince those who hold the purses that this task is worthwhile. Once the idea is demonstrated to be worthwhile, it is easily sold. But what they want to see is proof of how realistic is the subject. We should not just be trying to sell people ideas, something that is in the air. This is something that you should be extremely aware of - we must always keep under control the excitement, the opportunity, the expectations that are aroused; keep them realistic, keep them in gear with the real possibilities. If, as a result of these clinical trials, we show real positive promise, then I believe that we can move those at the centre and a budget will become available both for research and for clinical work.

I cannot make wild promises, however, that is the way things should be. I do hope that you have very rewarding deliberations in the course of this week and I look forward to a meeting in a year or two when we can share hopefully exciting new results.

Once more, thank you for coming, a warm welcome to you and rewarding deliberations.

WELCOME ADDRESS

H.H. van den Kroonenberg

Netherlands Energy Research Foundation (ECN) Petten
NL-1755 ZG Petten, The Netherlands

May I firstly wish you all welcome to ECN and to Petten. We are pleased and proud that you selected Petten as the venue for your workshop "Towards Clinical Trials of Glioma with BNCT". The primary objective of your co-operation is to install a clinical facility at the HFR in Petten. The owner of this High Flux Reactor is the JRC and therefore my dear friend Dr. Hondros, as the director of JRC in Petten, has the right to speak first, but of course this means that I have the last word.

Our share in this BNCT activity is evident not only because ECN personnel operate the reactor but also because our scientists are deeply involved in the development of the BNCT concept. Mr. Stecher-Rasmussen and Mr. Huiskamp will give presentations during this workshop. There is co-operation between JRC and ECN in many fields, and in most of these co-operations the joint interest is the HFR. For both of us it is vital to make clear to the community that the HFR is a very important and valuable machine for many purposes, serving the benefit of mankind. In this respect your research on BNCT and the role the HFR plays in it, is not only a matter of life and death for the patients but also for both our Institutes. That is why we appreciate your choice of Petten as the place where this clinical facility will be installed.

We frequently have to defend ourselves and our activities against strong forces in society. For example, the HFR makes use of High Enriched Uranium (HEU) as a fuel. There is strong pressure from certain political groups in society to force us to switch to Low Enriched Uranium (LEU). This would endanger our project. The reasoning behind this pressure is the belief that LEU is less vulnerable to proliferation. Presently in the US there are activists campaigning under the code name "Test Case Petten" who are trying to hold up a shipment of HEU for Petten. We do not fight these groups simply out of a desire to achieve our end, but we communicate with them because of our wish to convince them of our case.

A second battle in which we are involved is a very tedious dispute with animal welfare groups. We continuously have to convince them that we are very careful with animal tests and that we work well within the legal framework. The problem is that they do not accept the legal framework because they have a totally different objective. Ultimately you wish to help people. That is why all of you are involved in this programme. The animal welfare lobby is not against helping people but they want to achieve this end without making use of animals.

My message to you is to show that you carry out these investigations very carefully with a minimum of suffering to animals, even less than is legally permitted. This will provide us with valuable arguments in our discussions with these groups. I raise this issue to illustrate to you that in the background Dr. Hondros and myself make our own contribution to your programme.

I wish you a good and successful conference and hope that your stay in the Netherlands is memorable and enjoyable.

SAFETY AND EFFICACY IN

BORON NEUTRON CAPTURE THERAPY

Detlef Gabel

Department of Chemistry
University of Bremen
D(W)-2800 Bremen, F.R. Germany

INTRODUCTION

Boron neutron capture therapy (BNCT) is an attractive mode of radio-therapy. Its great potential lies in the fact that physiological differences between healthy and tumor cells can be utilized to enhance the dose to tumor cells selectively. This is achieved by administering, prior to the radiation treatment, a boron compound that can selectively accumulate or be retained in the desired target cells. Subsequently, the organ containing the tumor is irradiated with neutrons. Boron-10, by virtue of its high neutron capture cross section, will give rise to two densely ionizing particles through the $^{10}B(n,\alpha)^7Li$ neutron capture reaction.

The concept of BNCT is quite old. Locher[1] in 1936 was the first to propose the use of the $^{10}B(n,\alpha)^7Li$ neutron capture reaction for the treatment of tumors. In the past, clinical trials of treatment of high-grade glioma were conducted in the United States, during the 1950's and through 1961. They were abandoned as the treatment, as conducted then, gave rise to severe damage to healthy tissue. The reasons for the failure are now well understood. Contributing factors were a poor selectivity of the compounds utilized for tumor accumulation, a poor penetration of the neutron beam to deeper tissue, and overexposure of healthy tissue (see, e.g.,[2]).

Since 1968, Hatanaka[3] is treating high-grade gliomas, using $Na_2B_{12}H_{11}SH$ (BSH) as compound. As he is using a thermal neutron beam, the irradiation has to be done intraoperatively. Likewise, Mishima[4] is using a thermal beam to treat superficial melanoma lesions, with p-dihydroxyboryl phenylalanine (BPA) as compound. Both have reported successful treatments.

World-wide, a number of groups are working toward clinical implementation of BNCT, using epithermal neutron beams. These efforts, together with efforts in the preparation of new and improved compounds for delivery of boron to tumors, have started on a larger scale only around a decade ago, following a hiatus of almost 20 years after abandoning the last clinical trials.

In Europe, a Concerted Action "European Collaboration on Boron Neutron Capture Therapy" has as one of its aims to initiate clinical trials of treating high-grade glioma with the epithermal neutron beam of the High-Flux Reactor (HFR) and BSH as boron delivering compound in Petten, The Netherlands.

SAFETY AND EFFICACY OF BNCT

Epithermal neutron beams, which can be obtained from nuclear research reactors, have the advantage of delivering thermal neutrons also and predominantly to deeper-lying tissue. Thereby, tumors at depth can be treated without resorting to intraoperative techniques of radiation therapy. However, because of the past failure of BNCT, the initiation of renewed clinical trials of BNCT must be done with great care, and only after accumulating enough experience in model systems.

Two aspects need to be considered especially. These aspects concern the safety and the efficacy of BNCT. Failure to show that BNCT with a given compound can be administered safely would have great repercussions on the field as a whole, and possibly also on the use of research reactors for other purposes. Failure to show that BNCT can be an effective treatment would aprobably also inhibit its further development. It must be anticipated that these negative effects will prevail even when detailed reasons for lack of safety and/or efficacy can be given and corrections be suggested.

In radiotherapy, in contrast to chemotherapy, a clear-cut Phase I cannot be conducted easily, i.e. a study of whether a new treatment can be safe. Rather, safety and efficacy are the two sides of the same coin. Figure 1 illustrates this for a hypothetical case.

It is well established that low doses of radiation have little acute effects on single organs or parts of the body. Only once a "threshold dose" is surpassed, there is a steep increase of effect (i.e. damage in case of healthy tissue) with dose. For tumor control or tumor cure (i.e. the effect of radiation on a tumor) there is also a sharp increase with increasing dose, once a certain threshold is surpassed. As tumors are heterogeneous in size, the curve tends to be less steep than that for normal tissues. From these two curves, the effectiveness of a treatment modality can be derived as a difference between the observed benefit (i.e. tumor control) and risk (i.e. damage to healthy tissue). (It should, of course, be borne in mind that a representation of effectiveness as shown in Fig. 1 does not take into account the effects of the treatment on the patient as a whole, neither objectively nor as perceived subjectively. These factors need to be considered for a decision about how to treat a given patient with a given disease.)

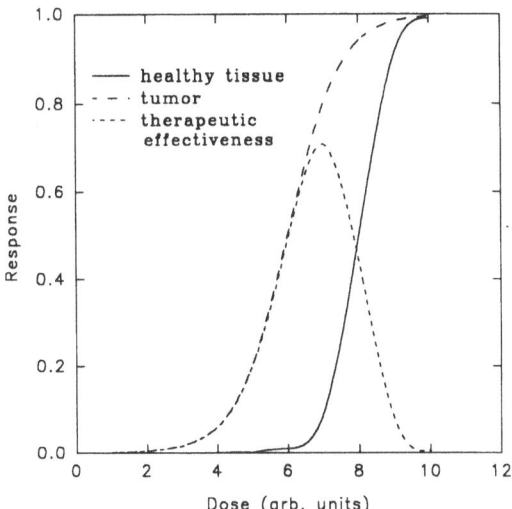

Figure 1. *Safety and efficacy of radiotherapy are interdependent. For the dose response of healthy tissue, a sigmoid curve is usually obtained. Also for tumor control, a sigmoid dose response curve is obtained. Because of factors such as number of cycling cells and the varying sizes of tumors, this curve tends to be less steep. Here, tumor response requires smaller doses than response of healthy tissue. The difference of the two curves would give some indication of the therapeutic effectiveness.*

Thus, it is easy to carry out any radiotherapy in a safe way, by choosing a low radiation dose. This will, however, not be beneficial for the patients. Only when the radiation doses approach the tolerance level of healthy tissue, can a benefit be achieved. Therefore, safety and efficacy cannot be separated from each other. It appears the aim of all clinical trials in radiotherapy to approach the safety limit from lower doses through dose escalation. In this process, when tumor response is also monitored, an indication of efficacy is obtained at the same time.

In BNCT, there are three aspects connected with safety and efficacy of a treatment. These three aspects concern

- the boron compound used for delivering boron to the tumor,
- the beam used for delivering neutrons to the tumor,
- the combination of compound and beam to deliver radiation dose to the tumor.

Safety and Efficacy Aspects of Boron Compounds

For BNCT to be successful, the compound employed must deliver boron to the tumor, and must deliver more boron to the tumor than to surrounding healthy tissue. It has been estimated that around 10-40 ppm boron-10 are needed in tissue to deliver most of the radiation dose from the $^{10}B(n,\alpha)^7Li$ reaction (see, e.g.,[5]). At the same time, concentration ratios exceeding 2:1, and extending to 10:1, between tumor and healthy tissue have been discussed as necessary for successful BNCT.

It is obvious that the compound and the route via which it is administered must not do undue harm to the patient. Acute and delayed toxic effects might have to be investigated. Lack of undue toxicity must be checked through a dose-escalation study, where increasing amounts of compound and thus boron are administered and vital functions of organs etc. are monitored. By doing this in patients scheduled for surgery of the type of tumor to be eventually treated, information can be obtained concerning tumor uptake and ratios of uptake between tumor and healthy tissue as a function of dose administered and time interval between administration and tissue removal. For subsequent radiation treatment, information will eventually be needed for each one of the healthy tissues exposed to the neutron beam.

Safety and Efficacy Aspects of Epithermal Neutron Beams

The use of epithermal neutrons beams for treatment of deeper-seated tumors presents two aspects. One aspect has to do mostly with safety, and concerns the whole-body dose associated with the beam delivery. The other aspect concerns the efficacy with which BNCT can be delivered from a given beam.

Components to the whole-body dose stem from insufficient shielding of the neutron source, and predominately from nuclear reactions (mostly hydrogen capture reactions) in collimator and target organ. These can be measured physically prior to irradiation. For objects of sufficient physical dimensions, such as larger animals and patients, the whole-body dose received might not pose a problem; for smaller animals, irradiation in an epithermal neutron beam will give rise to an unacceptably high whole-body dose. Neutron capture reactions in hydrogen, induced within the target volume, will also contribute to the whole-body dose. This component is unavoidable, but indirectly related to the boron capture dose.

The efficacy of BNCT depends greatly on the amount of additional dose which a given beam can deliver to a boron-enriched volume. There will always be a certain dose which is not associated with boron, and which also in an idealized situation can be avoided only partly. The part that cannot be avoided stems from the neutron capture in hydrogen in the target itself. The only practical way to reduce this dose is to increase the boron concentration in the tumor and hence the boron capture reaction dose.

However, for epithermal beams, technically unavoidable adventitious radiations doses are delivered by the beam itself. These doses consist of incident gamma photons (both from the reactor core, the surrounding material, and the neutron filter employed) and of fast neutrons (which could be filtered away only when seriously sacrificing the total neutron fluence). No experimentally determined limits of these two components, and especially the fast neutron component, are available. This partly due to the lack of knowledge about the RBE of the fast neutron and or the compound factor[6] of the boron capture components. It is obvious, however, that beams with smaller values for these two components, the incident gamma dose and the fast neutron dose, relative to the total neutron fluence would be better suited for NCT than those with higher values.

Safety and Efficacy Aspects of the Combination of a Particular Compound and a Particular Epithermal Neutron Beam

The total dose delivered to the different healthy organs (and the tumor) exposed to the beam stems from dose components associated with the beam and its interaction with boron-free tissue, and from the neutron capture reactions in boron. It is obvious that the total dose must not surpass the tolerance of healthy tissue.

Therefore, it must be determined experimentally which tissue is the critical (i.e. dose-limiting) healthy tissue. The nature of this tissue will depend critically on boron compound and concentration, field size of the beam, number of irradiation fields, and fractionation scheme employed. With the exception of boron compound and concentration, these dependencies are well-known in conventional radiotherapy.

For BNCT to operate at its best, much of the dose must obviously come from the neutron capture reaction in boron. It has been shown theoretically[7,8] and experimentally[9] that the effective dose delivered from boron will depend on the compound employed and its distribution within the target cells. At present, there is little evidence that the local efficacy of boron can be determined by methods other than a realistic model experiment. Especially, it appears very difficult to translate tolerance doses obtained for one particular compound to a different compound.

At this moment, it must therefore be concluded that safety and efficacy of BNCT cannot be separated. Moreover, there appears to be no possibility to predict the response of healthy tissue to exposure of neutron-induced radiation from different boron compounds. Due to the uncertainty with which an effective dose can be attributed to the $^{10}B(n,\alpha)^7Li$ reaction from a given compound (variations of one order of magnitude and more can be expected[8], each new compound for BNCT will require a dose-finding study. Only an empirical approach can be chosen for this.

SUMMARY OF THE PROBLEMS OF DETERMINING
SAFETY AND EFFICACY OF BNCT

From the physical and biological aspects of BNCT it must be concluded that safety and efficacy of a given treatment modality for BNCT can, for the time being, only be determined experimentally. This is due firstly to the fact that no single RBE factor can at present be assigned to the neutron dose delivered by broad-band filtered epithermal neutron beams. Secondly, and more important, is the fact that it cannot be predicted, which biological effect is associated with the physical dose delivered to healthy tissue and tumor from the neutron capture reaction of a boronated tumor seeking compound. Thus, an experimental study needs to determine both of these points. Due to the nature of epithermal beams and the whole-body dose to small animals associated with exposure to such beams, the needed parameters can be obtained only through the study of large animals or patients exposed to the treatment.

In patients especially, the approach must be dictated by the over-riding aspect of inflicting minimal harm. Therefore, a dose escalation study can proceed only very carefully, and in small increments. It must allow enough time for possible late tissue damage to develop before the dose can be escalated. For glioblastoma with its short median survival time of around eight months, the response of healthy brain to the treatment might therefore not be seen easily, as it would manifest itself only around the present median survival time. This would pertain both to a pure boron neutron capture therapy modality, and to a treatment modality where a locally well-defined dose of high-precision therapy is combined with a boost of BNCT. Dose escalation would therefore be possible only after enough patients had been treated and observed for around eight months to one year.

The European Collaboration on Boron Neutron Capture Therapy has therefore decided to choose the lesser of two evils. It will try to determine the effect of BNCT on healthy tissue in larger animals. By doing this, the initial experience from Gavin[10] can be incorporated into the design of this study, and the study can proceed much more rapidly. Results will be available at the end of the observation period of the last animal treated. The dose found to be safe in animals, reduced by a safety margin, will then serve as an estimate for the starting dose in patients. Careful dose escalation will then take place, allowing enough time to deduce the toxicity of the treatment to healthy tissue.

ACKNOWLEDGMENTS

The work of the European Collaboration on Boron Neutron Capture Therapy is supported by the Commission of the European Communities. I am grateful to my colleagues in the Collaboration for constructive and enthusiastic engagement in this work.

REFERENCES

1. Locher, G.L. Biological effects and therapeutic possibilities of neutrons. *Am. J. Roentgenol. Radium Ther.* 36, 1-13 (1936).
2. Slatkin, D.N. A history of boron neutron capture therapy of brain tumours. *Brain* 114, 1609-1629 (1991).
3. Hatanaka, H., ed. *Boron-Neutron Capture Therapy for Tumors.* Nishimura, Niigata (1986).
4. Mishima, Y.; Honda, C.; Ichihashi, M.; Obara, H.; Hiratsuka, J.; Fukuda, H.; Karashima, H.; Kobayashi, T.; Kanda, K.; Yoshino, K. Treatment of malignant melanoma by single thermal neutron capture therapy with melanoma-seeking 10B-compound. *Lancet II*, 388-389 (1989).
5. Fairchild, R.G.; Bond, V.P. Current status of ^{10}B-neutron capture therapy: Enhancement of tumor dose via beam filtration and dose rate, and the effects of these parameters on minimum boron content: A theoretical evaluation. *Int. J. Radiat. Oncol. Biol. Phys.* 11, 831-840 (1985).
6. Gahbauer, R.A.; Fairchild, R.G.; Goodman, J.H.; Blue, T.H. Can relative biological effectiveness be used for treatment planning in boron neutron capture therapy? In: *Tumor Response Monitoring and Treatment Planning.* Breit, A., ed., Springer, Berlin, in the press (1992).
7. Kobayashi, T.; Kanda, K. Analytical calculation of boron-10 dosage in cell nucleus for neutron capture therapy. *Radiat. Res.* 91, 77-94 (1982).
8. Gabel, D.; Foster, S.; Fairchild, R.G. The Monte Carlo simulation of the biological effect of the ^{10}B(n,α)^{7}Li reaction in cells and tissue and its implication for boron neutron capture therapy. *Radiat. Res.* 111, 14-25 (1987).
9. Coderre, J.A.; Makar, M.S.; Micca, P.L.; Nawrocky, M.M.; Joel, D.D.; Slatkin, D.N. Major compound-dependent variations of ^{10}B(n,α)^{7}Li radiation effectiveness for a rat gliosarcoma in vitro and in vivo. *Int. Workshop on Macro and Microdosimetry and Treatment Planning for Neutron Capture Therapy.* Plenum Press, New York, in the press (1992).
10. Gavin, P.R.; Kraft, S.L.; DeHaan, C.E.; Griebenow, M.L.; Moore, M.P. A large animal model for boron neutron capture therapy. In: *Progress in Neutron Capture Therapy for Cancer.* Allen, B.J.; Moore, D.E.; Harrington, B.V., eds., Plenum Press, New York, 479-484 (1992).

SOME ASPECTS OF BNCT RESEARCH AT THE

BROOKHAVEN NATIONAL LABORATORY

Darrel D. Joel[a], Vili Benary[b],
Robert M. Brugger[a], and Jeffrey A. Coderre[a]

[a]Medical Department
Brookhaven National Laboratory
Upton, N.Y. 11973, USA

[b]Tel-Aviv University
Tel-Aviv, Israel

INTRODUCTION

Prior to resumption of boron neutron capture therapy (BNCT) in patients at the Brookhaven National Laboratory (BNL), several preclinical issues need further evaluation. This report briefly addresses two of these issues, particularly characterization of the neutron beam at the Brookhaven Medical Research Reactor (BMRR) and boron compound selection.

MATERIALS AND METHODS

Dosimetric Characterization of BMRR Epithermal Neutron Beam

A series of radiation dose measurements were made using a cylindrical lucite phantom (15 cm diameter x 28 cm long) filled with water. The phantom was positioned with one end centered against the bismuth plate of the 'bare' 25 cm x 25 cm reactor port, with the axis of symmetry perpendicular to the port face. Bare and cadmium covered gold foils and thermoluminescent dosimeters (TLD 700) were placed at various depths in phantom and irradiated in the epithermal beam.

Boron Neutron Capture Therapy, Edited by D. Gable
and R. Moss, Plenum Press, New York, 1992

Brain Tumor Model

The transplantable gliosarcoma[1] used in this study was originally induced in a Fisher 344 rat by weekly i.v. injections of N-nitrosomethylurea[2]. Intracerebral tumors were initiated in Fisher 344 rats (Taconic Farms) that weighed 200-220 g by inoculating 1.0 μl of culture medium containing 10^4 cultured gliosarcoma cells into the left frontal lobe at a point 4-mm lateral to the midline and 1-mm anterior to the coronal suture. A 27-gauge needle was fitted with a Teflon collar to ensure cell injection 5 mm beneath the skull. This tumor does not produce blood-borne metastases. In preliminary studies, death ensued in 20 ± 3 (mean ± SD: n = 50) days from an expanding supratentorial tumor around the site of inoculation.

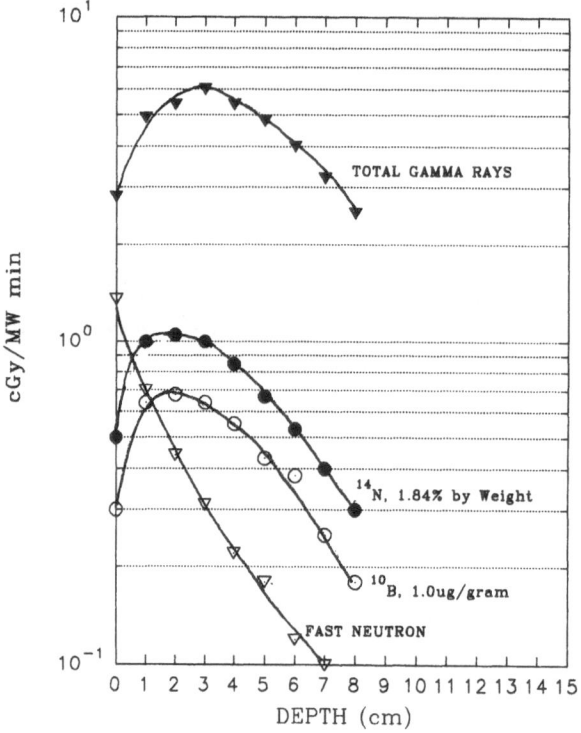

Figure 1. *Radiation dose rates in cGy/megawatt min from fast neutrons, the* $^{14}N(n,p)^{14}C$ *reaction, the* $^{10}B(n,\alpha)^7Li$ *reaction, and gamma rays as measured in a lucite cylinder phantom at the port face of the Brookhaven Medical Research Ractor. Points were fitted by hand.*

Table 1

*Calculated radiation dose rates to critical tissues of the head
assuming ^{10}B concentrations[7] approximating those obtained by
the administration of BPA to rats bearing gliosarcomas*

Site	Radiation	cGy/MW-min	RBE[2]	cGy·RBE/MW-min	% of dose
SKIN	$^{10}B(n,\alpha)^7Li$	1.2	2.3	2.8	(19.6%)
	Fast neutrons	1.8	4.4	7.9	(53.8%)
	$^{14}N(n,p)^{14}C$	0.5	2	1.0	(6.3%)
	Total gamma	2.9	1	2.9	(20.3%)
NORMAL	$^{10}B(n,\alpha)^7Li$	5.4	2.3	12.4	(60.8%)
BRAIN	Fast neutrons	0.2	4.4	0.9	(4.9%)
(4 cm deep)	$^{14}N(n,p)^{14}C$	0.9	2	1.8	(8.3%)
	Total gamma	5.3	1	5.3	(26%)
TUMOR	$^{10}B(n,\alpha)^7Li$	13.4	2.3	30.8	(84.6%)
(6 cm deep)	Fast neutron	0.1	4.4	0.4	(1.4%)
	$^{14}N(n,p)^{14}C$	0.5	2	1.0	(3%)
	Total gamma	4.0	1	4.0	(11%)

[1] ^{10}B concentrations = 10 $\mu g/g$ in blood and normal brain; 4 $\mu g/g$ in
skin; 40 $\mu g/g$ in tumor.
[2] RBEs: An RBE of 2.3 for the $^{10}B(n,\alpha)^7Li$ reaction was obtained
from the MIT Workshop on Neutron Beam Design[3,4]; RBE of 4.4 for
fast neutrons data obtained at BNL (unpublished results); an RBE of
2 for charged particles[5] from the $^{14}N(n,p)^{14}C$ reaction.

Clonogenic Cell Survival Following BNCT In Vivo

Rat brain irradiations were carried out on day 16 post-implantation. Immediately after BNCT, the rats were killed with an overdose of sodium pentobarbital anesthesia and the intracerebral gliosarcomas were removed aseptically, minced, and incubated with trypsin-EDTA (0.05% trypsin-0.53 mM EDTA in Hank's Balanced Salt Solution, without calcium or magnesium) for 30 min at 37°C. Fragments of tumor tissue were removed by centrifugation. Aliquots of the single-cell suspension obtained were diluted and plated for clonogenic colony-forming assay. For each tumor, two or three dilutions of the cell suspensions, with five replicate cell culture plates, were done. The plating (colony-forming) efficiency of cells from similarly-harvested, unirradiated tumors, was 55-65%. Survival fraction in the BNCT-treated tumors was normalized to the plating efficiency of cell suspensions from concurrently grown unirradiated tumors.

RESULTS AND DISCUSSION

Shown in Figure 1 are the dose rates in cGy/(MW·min) for fast neutrons, charged particles from the $^{14}N(n,p)^{14}C$ reaction (assuming 1.84% N by weight), the $^{10}B(n,\alpha)^{7}Li$ reaction assuming a ^{10}B concentration of 1.0 μg/gram tissue (brain), and total gamma rays. It can be seen from the $^{10}B(n,\alpha)^{7}Li$ and $^{14}N(n,p)^{14}C$ reactions that the thermal neutron flux is relatively low at the phantom face, peaks at 2 cm depth, and then decreases so that the flux at 6.5 cm depth is about equal to the incident flux. The dose rate from fast neutrons falls rapidly with increasing depth in the phantom.

Using these measurements, radiation doses to the scalp skin, normal brain tissue 4 cm beneath the scalp and brain tumor 6 cm beneath the scalp were calculated for an irradiation of 25 min at 3 MW assuming ^{10}B concentrations of 4 μg/g in skin, 10 μg/g in blood and normal brain, and 40 μg/g in tumor. These ^{10}B concentrations are reflective of those obtained with orally administered boronophenylalanine (BPA) in rats[6]. The results are listed in Table 1.

As expected, the largest dose to the skin is from fast neutrons. An RBE of 4.4 was used to calculate the fast neutron dose. This RBE may be too high, but is supported by preliminary studies with the 9L gliosarcoma (Coderre, unpublished data). The high value for the fast neutron RBE provides a "worst-case" scenario. Even with an RBE of 4.4, the total dose to the scalp would be about 11 Gy (assuming a 75 MW·min exposure) which is below that which will result in skin necrosis. If, on the other hand, ^{10}B levels in the skin were significantly higher, i.e., >20 μg/g, serious radiation damage may result.

This raises the question of whether normal skin tolerance studies should be done in swine, which is believed by nearly all investigative dermatologists to be the best model for predicting effects in humans. As with normal brain, skin tolerance associated with BNCT will depend upon tissue distribution of the boron transport agent used.

The dose to a 6 cm deep tumor will be largely from the $^{10}B(n,\alpha)^{7}Li$ reaction. With a 75 MW·min exposure ($\approx 3 \cdot 10^{12}$ thermal neutrons cm^{-2}), a total dose of about 27 Gy·RBE would be achieved. Under these same conditions, the calculated dose to normal brain 4 cm deep would be about 15 Gy·RBE. A dose of 15 Gy·RBE to the cerebral cortex may be tolerated, however, similar doses to the basal ganglia are likely to result in unacceptable brain damage[5].

Boron Compound Selection

To our knowledge, the only compounds that have been thoroughly tested in animal tumor models have been the sulfhydryl boranes, monomer (BSH) and dimer (BSSB), and boronophenylalanine (BPA). Of these compounds, we

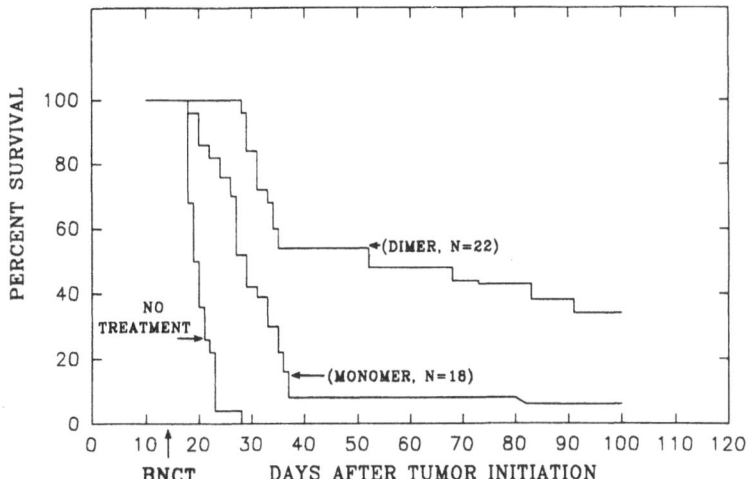

Figure 2. *Effects of BNCT on the survival of rats bearing intracerebral gliosarcomas. Rats were infused with either BSH or BSSB and exposed to 5 megawatt min reactor radiations ($\approx 2 \cdot 10^{12}$ thermal neutrons cm^{-2} to tumor) at a time when the blood ^{10}B levels were 34.5 $\pm$ 2.4 µg/ml and 34.9 $\pm$ 5.5 µg/ml, respectively.*

have been able to achieve 50% long-term (i.e., greater than 1 year) survival in rats bearing gliosarcomas with BPA or BSSB[6,7], but not with BSH. The efficacy of BNCT on the survival of rats bearing gliosarcomas was compared in two groups of rats infused with either BSH or BSSB. The average blood ^{10}B concentration at the time of exposure to reactor radiations was the same in both groups, i.e., 34.5 $\pm$ 2.4 µg/ml and 34.9 $\pm$ 5.5 µg/ml, respectively. Previous studies[8] would indicate that the brain tumor boron concentration was about 25 ppm in both groups of rats since the tumor-to-blood ratio has been shown to be the same for both of these compounds.

As shown in Figure 2, rats infused with BSH and exposed to 5 MW-min reactor radiation had a median survival of 30 days post-implantation with only 1 of 18 rats surviving more than 110 days (cure). In contrast, the median survival of BNCT-treated rats infused with BSSB was 60 days with 32% (7 of 22) long-term survival. Since the average tumor ^{10}B concentration was the same in both groups of rats, it follows that the different survival times in the two groups were most likely due to dissimilar, compound-dependent microdistributions of boron.

To further test the efficacy of these boron compounds, rats bearing intracerebral gliosarcomas were administered either BSH, BSSB or BPA at a rate which would yield on the average 30-35 µg ^{10}B/gram tumor. Rats were exposed to 5 MW-min reactor radiation, euthanized, and the tumor-clonogenic cell survival determined. The results are listed in Table 2. The clonogenic cell survival in tumors removed from rats administered BPA was very low; in some cases no colonies were detected with the cell dilutions used. In contrast, numerous clonogenic cells survived BSH- or BSSB-based BNCT, with BSH being significantly less effective than BSSB.

Table 2

The average number of colony-forming units derived in vitro *from cerebral gliosarcomas treated with BNCT* in vivo[1]

Boron Compound Infused	colony-forming cells per 10,000 cells plated[2]
BPA	3 ± 3
BSSB (dimer)	230 ± 60
BSH (monomer)	370 ± 100
x-rays (20 Gy)[3]	30 ± 10

[1] The average tumor ^{10}B concentration for each compound was ≈ 30 μg/gram and the thermal neutron fluence was $2 \cdot 10^{12}$ $n_{th}cm^{-2}$.

[2] Survival was normalized to the plating efficiency from unirradiated tumors.

[3] Rats were exposed to x-irradiation using a 250 kVp x-ray generator at a dose rate of about 3.75 Gy/min (30 mA, 0.5-mm Cu, and 1.0-mm Al filtration).

Both BPA and BSSB have been shown to be equally effective boron agents for controlling rat gliosarcomas by BNCT[6,7]. The lack of correlation between *in vivo* efficacy and clonogenic cell survival may be related to different microdistributions of ^{10}B at the tissue and/or cellular level. BPA apparently enters tumor cell rather uniformly, resulting in direct killing or sterilization of nearly all potential clonogenic cells. BSSB, on the other hand, may concentrate in tumor vasculature to a greater extent than BPA, resulting in indirect or delayed cell killing through diminished blood supply to the tumor. BSH may be similar to BSSB in distribution, but may not concentrate to an effective level in either tumor cells or vasculature.

CONCLUSIONS

The epithermal beam at the BMRR is presently adequate for the treatment of ocular melanoma and brain tumors up to 6 cm in depth with BNCT. For deep-seated brain tumors, it may be necessary to do bilateral exposures. Although the largest fraction of the radiation dose to the skin may be from fast neutrons, it does not appear that skin will be the limiting tissue when patients are exposed to reactor radiations for time periods needed for effective thermal neutron fluences at tumor depth unless there is a high concentration of ^{10}B in critical skin tissues.

Selection of the most effective boron compound is difficult. BPA has been shown in animal models to be effective in controlling experimental brain tumors *in vivo* and was clearly the most effective compound tested in the clonogenic survival assay. The levels of ^{10}B in normal brain following BPA administration are, however, quite high (≈ 10 $\mu g/g$). BSSB is also effective *in vivo*, but the hepatotoxicity associated with this compound may render it unacceptable for administration to humans. BSH is less toxic than BSSB, but also less effective in controlling tumors *in vivo*. Clearly, further compound development will be critical to the success of BNCT as an acceptable therapy for malignant brain tumors.

ACKNOWLEDGEMENTS

Research supported in part by the U.S. Department of Energy under Contract DE-AC02-76CH00016.

REFERENCES

1. Barker, M.; Hoshino, T.; Gurcay, O.; Wilson, C.B.; Nielsen, S.L.; Downie, R.; Eliason, J. Development of an animal brain tumor model and its response to therapy with 1,3-bis(2-chloroethyl)-1-nitrosourea. *Cancer Res.* 33, 976-986 (1973).
2. Schmidek, H.H.; Nielson, S.L.; Schiller, A.L.; Messer, J. Morphological studies of rat brain tumors induced by N-nitrosomethylurea. *J. Neurosurg.* 34, 335-340 (1971).
3. Slatkin, D.N.; Kalef-Ezra, J.A.; Saraf, S.K.; Joel, D.D. Beam modification assembly for experimental neutron capture therapy of brain tumors. In: *Neutron Beam Design, Development and Performance for Neutron Capture Therapy, Basic Life Sciences* 54. Harling, O.K.; Bernard, J.A.; Zamenhof, R.G., eds., Plenum Press, New York, 317-320 (1990).
4. Gabel, D.; Fairchild, R.G.; Larsson, B.; Börner, H.G. The relative biological effectiveness in V79 Chinese hamster cells of the neutron capture reactions in boron and nitrgen. *Radiat. Res.* 98, 307-316 (1984).

5.   Slatkin, D.N. A history of boron neutron capture therapy of brain
     tumors. Postulation of a brain radiation dose tolerance limit. *Brain*
     114, 1609-1629 (1991).
6.   Coderre, J.A.; Joel, D.D.; Micca, P.L.; Nawrocky, M.M.; Slatkin, D.N.
     Control of intracerebral gliosarcomas in rats by boron neutron cap-
     ture therapy with p-boronophenylalanine. *Radiat. Res.* 129, 290-296
     (1992).
7.   Joel, D.D.; Fairchild, R.G.; Laissue, J.A.; Saraf, S.K.; Kalef-Ezra, J.A.;
     Slatkin, D.N. Boron neutron capture therapy of intracerebral rat
     gliosarcomas. *Proc. Natl. Acad. Sci. USA* 87, 9808-9812, (1990).
8.   Joel, D.; Slatkin, D.; Fairchild, R.; Micca. P.; Nawrocky, M. Pharma-
     cokinetics and tissue distribution of the sulfhydryl boranes (monomer
     and dimer) in blioma-bearing rats. *Strahlenther. Onkol.* 165, 167-170
     (1989).

INEL BNCT PROGRAM DIRECTIONS WITH RESPECT TO

CLINICAL TRIALS OF BNCT

Ronald V. Dorn III

INEL BNCT Program
Idaho National Engineering Laboratory
EG&G Idaho, Inc.
Idaho Falls, ID 83415, USA

INTRODUCTION

The Idaho National Engineering Laboratory's Boron Neutron Capture Therapy (INEL/BNCT) Program was initially established in 1984 to provide a new improved, effective treatment for Glioblastoma Multiforme. To address this need, the Program was organized with four goals in mind: (1) develop BNCT as a more effective brain tumor treatment, (2) expand BNCT application to other tumor types, (3) develop better boron compounds, and (4) deploy the technology to serve more patients.

Many of the technology requirements necessary for the achievement of Goal 1 (institution for BNCT clinical trials) have been developed during the first years of the Program. This paper will review the logical progression of technology development within the INEL BNCT Program since 1984 and our vision of progress towards clinical trials of BNCT in the near term.

TECHNOLOGY DEVELOPMENT

At the onset of the INEL BNCT Program, the need was determined to develop new technologies and to enlarge upon the preclinical database to allow safe and effective human treatments. Toward this end, the following research tasks were established:

1. Large animal model system to adequately test epithermal BNCT *in-vivo*. The spontaneously-occurring brain tumor dog was chosen. This model has been successfully developed with an operational recruit-

ment network. Canine intravenous $Na_2B_{12}H_{11}SH$ (borocaptate sodium or BSH) pharmacokinetics have been established, as well as tumor- and normal-tissue concentrations and plasmapheresis effects. Fig. 1 shows the compiled data from 24 dogs receiving 55 g boron/kg (includes dogs with other central nervous system lesions and control dogs).

2. An epithermal-neutron beam "filter" was designed, installed, and characterized at the Brookhaven Medical Research Reactor (BMRR) to facilitate *in-vivo* epithermal BNCT animal studies.

3. A three-dimensional BNCT dose-distribution computational tool has been developed, validated, and is now operational.

4. Initial canine dose-tolerance studies have been completed demonstrating:

    a. $0 \, {}^{10}B$ (10 Gy < scalp tolerance < 15 Gy)
    b.   25 ppm ${}^{10}B$ intrabrain tolerance of 21 ± 2 Gy
    c.   50 ppm ${}^{10}B$ intrabrain tolerance of 28 ± 2 Gy
    d.   Large vascular endothelium ${}^{10}B$ capture dose sparing
    e.   High RBE of fast-neutron beam contaminate (RBE $\approx$ 4-6)

5. Several brain-tumor dogs have been safely treated with impressive initial tumor responses and some long-term tumor controls.

6. Based on the initial dose-tolerance studies and brain-tumor dogs results, 100-ppm ${}^{10}B$ studies are now underway with anticipated improved tolerance and tumor responses.

7. Inductively coupled plasma-atomic emission spectroscopy (ICP-AES) has been developed as the reference standard procedure for boron analysis in biological samples.

8. Boron compound purity analysis procedures have been developed and transferred to boron drug manufacturers and in support of the U.S. Food and Drug Administration BSH Investigational New Drug Exemption (IND) application.

9. Intra- and intercellular boron localization and quantification by Secondary Ion Mass Spectroscopy (SIMS) has been demonstrated and is under refinement.

10. Noninvasive tumor-boron measurement by magnetic resonance imaging spectroscopy (MRI/S) has been demonstrated and used in support of canine tumor pharmacokinetic studies. Active refinement is underway for application in preclinical and clinical studies.

11. The INEL BNCT Program has expanded into metastatic malignant melanoma research and boron compound development.

Status of Goal 1

With the development of this technology, the status of the INEL BNCT Program's Goal 1 (to provide more effective brain tumor treatment by) is: (1)

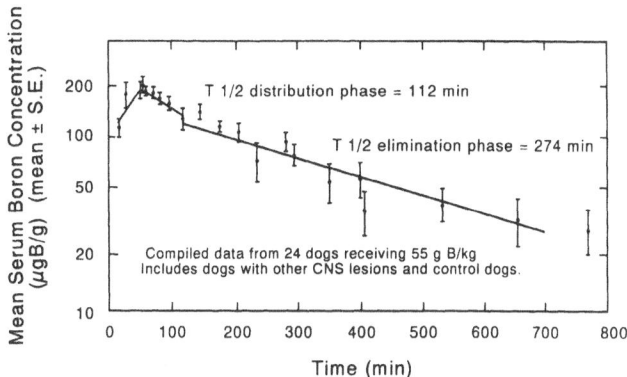

Figure 1. *Mean serum concentration (± S.E.) for tumor-bearing dogs.*

required technology and tools to pursue this goal have been developed, (2) BSH dosage optimization and efficacy studies in the canine model system have been conducted, (3) studies in support of optimization of neutron sources have been completed, and (4) further assessment of BNCT radiation biology and radiation response continues.

With this firm technological and scientific database established, the INEL BNCT Program direction and plans in further pursuit of Goal 1 include completion of the BNCT radiation biology studies in small and large animal models, development of an appropriate radiation source for human BNCT at either a reactor or, possibly, an accelerator source, continuation of human pharmacokinetic studies, and initiation of human clinical trials.

Clinical Trials

Based on the technology and scientific progress within Goal 1 (outlined above), the following prerequisites and INEL BNCT Program structure (Fig. 2) have been established in anticipation of human clinical trials of BNCT:

1. There must be a basic understanding of the radiobiology of BNCT at the cellular, small animal, and large animal levels.
2. There must be further elucidation of normal tissue response to BNCT. This is currently being addressed with optimization of BSH administration, as well as optimization of the radiation source (INEL BNCT Program: Project 2, Task 1).
3. There must be further definition of physiological response to include the radiation pathophysiology of brain BNCT injury vs. gamma injury, fractionation effects, delayed injury prediction, and injury prevention (INEL BNCT Program: Project 2, Task 3).

4.   Human pharmacokinetic studies with the boron compound must be pursued with attention to different albumin binding characteristics, optimum dose, and toxicity (INEL BNCT Program: Project 3, Task 1).

5.   Response of spontaneous and induced tumors must be further analyzed at various levels of complexity, *in-vitro* and *in-vivo* (INEL BNCT Program: Project 2, Tasks 1 and 3).

6.   Ideally, noninvasive boron localization and concentration determination for use in patient screening and dosimetry (MRI/S) should be operational (INEL BNCT Program, Project 1, Task 2).

7.   Drug interaction and toxicity studies involving the boron compound should be undertaken (INEL BNCT Program, Project 4).

8.   As a clinical trial prerequisite, additional technology and tool development should include further studies on boron concentration at the cellular and tissue levels, refinement of mixed-beam dosimetry, operational three-dimensional mixed-radiation treatment planning and microdosimetry, and, finally, optimization of the radiation source to decrease fast-neutron contamination.

These clinical trial prerequisites are currently under investigation and development within the INEL BNCT Program and other world-wide BNCT research programs. The primary overall thrust of these prerequisites is to ensure, as much as possible, the safety of BNCT prior to its human application, and secondly, its effectiveness.

A survey of the current state-of-the-art in BNCT, as well as these prerequisites, indicate that U.S. clinical trials should be ready to begin sometime within the 1993 to 1995 time frame. INEL BNCT Program participants feel strongly that, based on data obtained to-date, these clinical trials within the

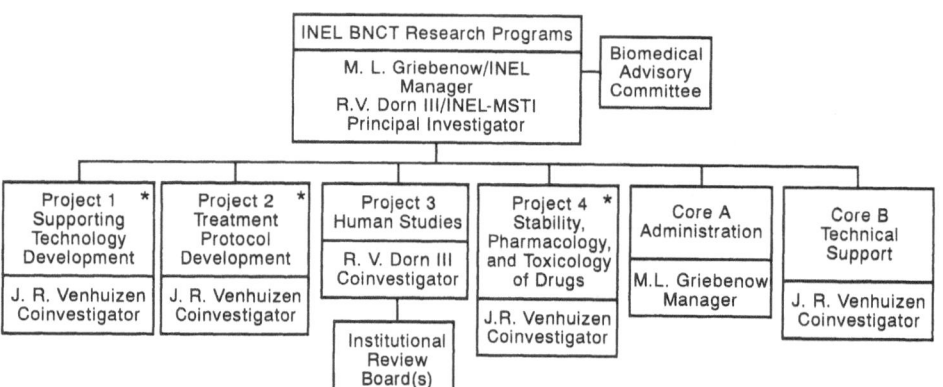

Figure 2. *INEL BNCT Program structure.*
* Accountable to Institutional Animal Care and Use Committee

INEL BNCT Program should be directed toward brain tumors (Glioblastoma and brain metastases) using the BSH compound. Our vision of the radiation source involves either an appropriately-modified reactor source or perhaps an accelerator source. Clinical trials should be conducted as Phase I, then Phase I/II addressing single dose, split dose, and limited fractionation treatment schedules. Endpoints of these initial clinical trials should be safety first, dose escalation second, and, finally, tumor response. The trials should be local/-regional followed by national in scope, appropriately involving a national cooperative research group such as the Radiation Therapy Oncology Group (RTOG).

INEL BNCT Program Goals 2, 3, and 4

Once clinical trials have begun within the framework outlined above, Goals 2, 3, and 4, initially established within the INEL BNCT Program, can then be addressed. That is, with ongoing clinical trials, further research can take place at the preclinical and clinical levels to investigate expansion of BNCT treatment to other tumor types, along with the appropriate development of improved boron compounds and deployment of BNCT technology to serve larger numbers of patients at regional treatment centers.

CONCLUSIONS

INEL BNCT Program participants feel strongly that it is inappropriate to withhold BNCT clinical trials indefinitely, while waiting for the "perfect" boron compound. Once BNCT, with currently-available compounds, has been determined to be safe and potentially effective, clinical trials should proceed in view of the lethal nature of these tumors and the possibility of improved quality of life in these patients (due to decreased length of treatment time compared to conventional treatment, etc.). A strong historical precedent exists within the fields of both conventional radiation therapy and systemic chemotherapy for ongoing treatment optimization after the initial establishment of treatment technique, safety, and response.

ACKNOWLEDGMENT

This study was performed under the auspices of the U.S. Department of Energy, Office of Energy Research, under DOE Field Office, Idaho, Contract No. DE-AC07-76ID01570.*

---

* The submitted publication has been authored by a subcontractor of the U.S. Government under DOE Contract No. DE-AC07-76ID01570. Accordingly, the U.S. Government retains a nonexclusive, royalty-free license to publish or reproduce the published form of this contribution, or allow others to do so, for U.S. Government purposes.

THE DEPARTMENT OF ENERGY RESEARCH PROGRAM

IN BORON NEUTRON CAPTURE THERAPY

Robert W. Wood and **Donald W. Cole Jr.**

U.S. Department of Energy
Washington, DC 20585, USA

For the past four decades the Department of Energy and its predecessor agencies, the Energy Research and Development Administration and the Atomic Energy Commission, have sustained a research program on Boron Neutron Capture Therapy. In recent years the Department's research program has expanded in response to the promise of several new boron localizing agents and the development of improved epithermal neutron beam delivery capabilities. Our plans for future activities encompass a number of specific areas for research emphasis.

The development of accelerator-based neutron sources is of interest as a preferable method for producing epithermal neutron beams of sufficient power. Such a source would permit the use of BNCT in many hospital settings, rather than just at nuclear reactors. Design studies are now being conducted for accelerator-based epithermal neutron beams. Future efforts will be required to assemble and test an accelerator and neutron beam moderator assembly. However, the need for such developments is contingent upon the demonstration that BNCT will be useful in cancer treatment. Similarly, it would be premature to convert a reactor until efficacious boronated compounds exist.

At the present time, a major significant need in the BNCT area is in the area of boron compound synthesis, delivery, and evaluation to see if compounds can be found that concentrate in the tumor so that normal cells will be relatively unaffected when exposed to neutrons. Existing compound development and delivery programs need to be explored for new systems. In addition, resources need to be invested in the creation and operation of new evaluation centers that are fully equipped with spectrometers, and a suitable neutron source, coupled to the ability to rapidly carry out *in vitro* and *in vivo* boron localization and toxicity studies with new compounds and delivery methods. Experimental means to study the cellular uptake of new compounds

Boron Neutron Capture Therapy, Edited by D. Gable
and R. Moss, Plenum Press, New York, 1992

need to be established for evaluation of new boron compounds at the cellular level and compared to a boric acid standard. The most promising compounds would be used in actual irradtion experiments with animals combined with an appropriate neutron beam. Rapid progress in compound synthesis can be enhanced with the guidance of experimental animal studies performed currently.

Modern drug-design theories have not been extensively applied to the syntheses of BNCT agents. Using the knowledge acquired in the chemotherapy arena in recent years, there are a number of "fourth generation" BNCT agents which can be envisaged and require synthesis and evaluation. These agents include boron analogues of naturally occurring small molecules such as amino acids, porphyrins, nucleosides, low density lipoproteins, and a variety of amino bases as well as macromolecules which could include tumor specific antibodies. These opportunities for development of more efficacious boronated compounds need to be pursued more aggressively in the future.

An enormous variety of chemical approaches to boron delivery are possible and are limited only by the imagination of the chemist and by the resources available. A coordinated national program has the potential to provide desperately needed resources to bring to research laboratories exciting, currently available compounds and to create new and even better agents.

An important determinant of BNCT must be the microdistribution of boron-10 in tumors, in endothelium, and in blood. Experimental studies of the most complex problems of BNCT - improving boron distribution and neutron penetration - are most useful at this stage of research and development. The effective radiation dose to tumors from the interaction of thermal neutrons and boron-10 is not necessarily proportional to the product of the incident flux and the average concentration of boron in the tumor integrated over the duration of the irradiation. An important dosimetric consideration is whether most of the boron is extracellular or bound to the outer cell membrane, to the cytoplasm, or to the neclei of tumor cells or in the blood stream.

Current *in vivo* techniques center on track-etch methodologies which utilize autoradiography of tissue samples containing boron-10 after they have been bombarded with neutrons. The content of tissue samples can also be determined using digestion techniques followed by spectroscopic examination of the digestate. These are valuable tools, but they cannot be used to determine the boron concentration within cells or evaluate the inter/intracellular location of the boron. New techniques will need to be developed, such as NMR-microscopy, Resonance Ionization Spectroscopy, Element-Specific X-Ray microscopy, and others yet to be identified.

Demonstrations of biological feasibility of a proposed cancer treatment in appropriate test systems (i.e., cell culture and/or animal tumor models) are considered by oncologists to be mandatory before clinical trials. In the initial development of a new class of boronated compounds, it is important to assess tumor cell affinity and effective subcellular localization as indicated in the response of cell cultures to the $^{10}B(n,\alpha)^7Li$ reaction. This is justifiable because various classes of boron compounds (boronated nucleosides, monoclonal antibodies, liposomes, and low density lipoproteins) cause boron to be transported between cell membranes, various cytoplasmic compartments and nuclei via different, usually unknown or poorly understood, biochemical mechanisms. Furthermore, the details of subcellular boron-10 partitioning, expecially as nuclear boron-10 transport and retention are concerned, are critical determinants of BNCT feasibility. *In vitro* testing may be the only practical route of testing possible for compounds that are available initially in amounts under ≈ 300 mg because of the expense of the enriched boron-10 compound.

Biological feasibility of boronated compounds is evaluated in small animals by irradiating 1.1 - 1.5 cm diameter murine (mouse and rat) tumors carried subcutaneously in the thigh. Thermal neutron beams from the Brookhaven Medical Research Reactor (BMRR) are used for this *in vivo* technique. To evaluate the effectiveness of boron-10, results from reactor radiations with neutrons alone are compared with results of neutron irradiations of boronated mice. The program needs to be expanded at the BMRR and at two or three other sites to facilitate boron compound and evaluation.

The strategy in cancer radiotherapy is to irradiate normal tissues to their tolerance levels in the effort to deliver as much dose to the tumor as possible. A detailed evaluation of the distribution of each beam component, along with the relative biological effectiveness appropriate for the beam component and fractionated regimen employed, is required to accurately predict normal tissue response. Computer techniques allow the computation of three-dimensional dose distributions, and accurate isodose charts. These procedures will be required for sophisticated BNCT clinial trials that can be compared with other forms of radiation therapy of brain tumors.

Normal tissue tolerance to an epithermal beam needs further evaluation. Measurements will include the determination of tolerance dose to tissues with and without boron. Presently, studies have demonstrated a tolerance of the normal central nervous system in long-term surviving rats successfully treated for malignant transplanted gliosarcoma. In the future, similar tolerance studies in non-tumor bearing dogs are needed. The primary goal is to quantitatively relate normal tissue tolerance studies in animals to human therapy which is needed prior to possible clinical trials.

Boron compounds can be used for cancer treatments other than those with epithermal neutrons. For instance, the possibility of enhancing the effectiveness of fast neutrons for glioblastoma therapy offers a near-term treatment procedure. The fast neutrons exhibit the same skin-sparing effect as epithermal neutrons and could provide an enhancement of about 30 percent which could significantly increase the probability of tumor control using existing technology. Research efforts to obtain cell survival curves for boron compounds suitable for fast neutron therapy and their approval by the FDA could result in early clinical trials.

CURRENT OVERVIEW ON THE

APPROACH OF CLINICAL TRIALS AT PETTEN

Raymond L. Moss

Commission of the European Communities
Institute for Advanced Materials
Joint Research Centre
NL-1755 ZG Petten, The Netherlands

INTRODUCTION

In August 1990, the main components of the filtered (epithermal) neutron beam were installed in the beam tube HB11 of the High Flux Reactor (HFR) at Petten[1]. Thereafter followed a prolonged 9 months period of testing and commissioning before it was finally possible in June this year to open the beam for the first time at full reactor power. The first measurements to characterise the neutron beam were made, followed shortly by some initial cell culture irradiation experiments, and the start of the *in vivo* healthy tissue tolerance studies using the canine brain.

Over the previous year, development work also continued and progressed in the areas of boron detection techniques, dosimetry and treatment planning schemes. In addition, some initial thoughts on the procedures and protocol for the start of clinical trials, and the installation of a patient treatment room, have been developed.

In the sections below, a brief overview of the HFR, the beam tube HB11 and the present configuration of the irradiation room are given. The following sections review superficially the aims and progress of the experimental programme. More detailed information on each topic are presented by colleagues at this Workshop. In the final section, overviews are given on the clinical set-up as planned for clinical trials, on the extension of this facility to beyond clinical trials and on the expected, continued lines of research in BNCT at Petten.

The work at Petten forms an integral part of the European Collaboration group's activities on BNCT[2]. In particular, the direct activities at Petten are shared and coordinated amongst colleagues at the Joint Research Centre

(JRC, Petten), the Netherlands Energy Research Foundation (ECN, Petten) and the Netherlands Cancer Institute (NKI, Amsterdam). The work of the European group and the project at Petten is primarily funded through the European Commission's Medical and Health Research Programme in Brussels.

THE HIGH FLUX REACTOR AND BEAM TUBE HB11

The High Flux Materials Testing Reactor (HFR)[3] at Petten is owned and managed by the Commission of the European Communities. For the last 25 years, the HFR has been utilised primarily for the irradiation testing of materials and nuclear fuels for the European civil nuclear power programme. The reactor operates at 45 MW and is cooled and moderated by light water. The standard core configuration consists of 33 fuel assemblies, 6 control rods, 16 beryllium reflector elements and 17 free positions for placing experimental facilities. The reactor is also equipped with 12 horizontal beam tubes, used mainly for nuclear physics and solid state physics research.

For BNCT applications, a sufficient and adequate flux of neutrons can only be acquired from a high flux reactor and then only through a large aperture beam tube, that itself should preferably face a large source area of the reactor. The beam tube arrangement, HB11/12 at the north side of the HFR reactor vessel, satisfies these physical requirements[4]. In addition, the exit

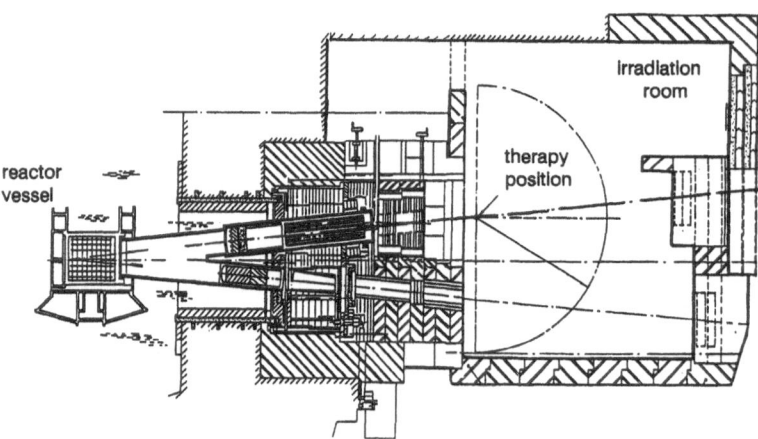

Figure 1. *Overview of the planned lay-out of the Petten treatment room, including the reactor vessel and the HB11 beam tube (upper tube in diagram) configuration.*

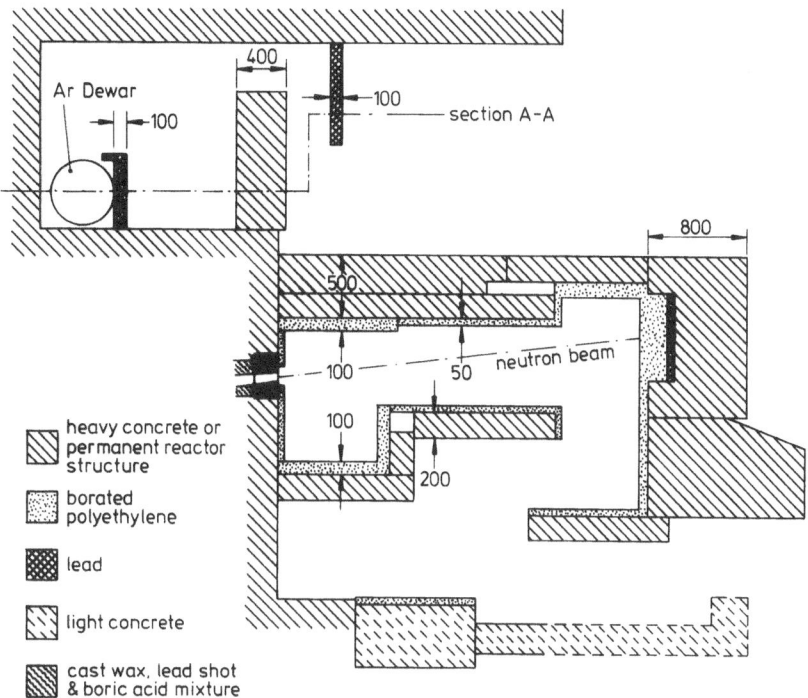

Figure 2. *Present set-up of the HB11 experimental irradiation room.*

side of the beam can accommodate a large working area for developing an irradiation room of sufficient size for performing experiments and for the eventual treatment of patients positioned in lateral recumbence, see Fig. 1. Furthermore, the facility faces the reactor building's emergency exit, that in effect gives unhindered access from the outside.

Following 2 years of pre-studies and measurements to confirm the favourable nuclear characteristics of the proposed beam tube, the design for the required filter configuration to produce an epithermal neutron beam at HB11 was completed. The design goals for the beam for BNCT applications were specified by the Project Management group of the European CA group, as follows:

neutron flux $> 1.0 \cdot 10^9$ neutrons / cm^2 s
(at the therapy position)

average neutron dose $< 8.0 \cdot 10^{-11}$ rads cm^2 / neutron
(or mean neutron energy < 8.0 keV)

and

gamma dose rate < 50 cGy / hr.

The requirements were chosen to enable a treatment to be completed in a reasonable time (< 1 hour), probably in 6 fractions ($\approx$ 10 min. each). The average neutron dose can be re-interpreted using the ICRU neutron dose function (i.e. 10^{-11} rads cm^2/neutron $\approx$ 1 keV) to produce a mean energy in keV. This is considered a useful indication of the quality of the beam and reflects the fast neutron contamination in the beam. The incident gamma dose rate must be sufficiently small as not to dominate the total (incident + induced) gamma dose rate within the target volume.

The design of the facility was performed jointly between the Petten group and groups at AEA Harwell (UK) and JRC Ispra (Italy)[5]. The calculations were performed using the Monte Carlo code, MCNP[6]. The following combination of filter materials to produce the optimal beam characteristics was chosen: 15 cm Al; 5 cm S; 1 cm Ti; 0.1 cm Cd; and 150 cm liquid Ar.

Installation of the various components, including the above filter materials, a new main beam shutter, an emergency beam shutter and new shielding blocks, was completed in August 1990. A provisional irradiation room, fully shielded, was built around the beam port, see Fig. 2. This small room is sufficient for carrying out the planned experimental programme which aims to determine many of the pre-conditions for clinical trials. For clinical trials with patients, it will be necessary to enlarge this room.

COMMISSIONING OF THE FACILITY

Despite the advantages of utilising a high flux nuclear facility, it is nevertheless necessary that certain stringent safety criteria must be satisfied before experimental work may be performed. These constraints, as laid down by safety committees guided by nuclear installation rules, meant that many extra safety provisions, such as back-up systems, emergency procedures etc. all had to be documented and demonstrated before permission to open the beam at full reactor power could be obtained. In addition, the uncertainties attached to using liquid argon, gave unprecedented attention to this facility from the safety committees. Whilst this may be prudent, it gave considerable delays to the programme. Nevertheless, after almost 9 months of improving and testing the system, it was finally possible in June of this year, to open the main beam shutter at full reactor power. It should be mentioned that on earlier occasions, conditional permission was given to operate the system at low reactor power, thus gaining insight into the operation of the system by performing some initial beam measurements. Prior to the opening of the beam at full reactor power, an extensive amount of measurements was performed by the group from Harwell, to measure the flux and spectrum characteristics at the higher end of the neutron energy spectrum[7].

FROM NOW TO CLINICAL TRIALS

With the beam available at full reactor power, the programme of work towards the realisation of clinical trials, as defined by the Petten BNCT Group, has started. The planned activities concentrate on determining the radiobiological limits associated with the irradiation of healthy tissue with an epithermal neutron beam. An outline of the programme is summarised as follows :

- Characterisation of the unperturbed beam and evaluation of the neutron and gamma fields in plastic phantoms,
- cell culture experiments with V79 Chinese hamster cells, looking at different concentrations of ^{10}B (boric acid, BSH, BPA) and at different depths inside a plastic phantom filled with tissue equivalent liquid,
- healthy tissue tolerance studies on the canine brain, to determine the limiting dose of the epithermal neutron beam to the healthy brain,
- development of a treatment planning scheme, coupled with phantom experiments and 2D- and 3D-graphics routines,
- design of the treatment room for clinical trials.

Nuclear Characterisation of the Beam

In July, a collaborative effort between INEL USA[8], AEA Harwell UK[7], ECN[9] and JRC Petten carried out the following measurements using standard dosimetry techniques : stacked resonance-energy activation foils for measuring the neutron spectrum in the epithermal energy range; energy response threshold foils; filtered response threshold foils; proton recoil proportional counters for measuring the spectrum in the fast neutron energy range; and thermoluminescence detectors (TLD) for measuring the gamma field. The resultant exercise[8] gave a total neutron flux of $4.9 \cdot 10^8$ cm^{-2} s^{-1}; an epithermal neutron flux of $3.8 \cdot 10^8$ cm^{-2} s^{-1} ; a fast neutron dose rate of 1.5 Gy/hr and a gamma dose rate of 1.0 Gy/hr. The mean neutron energy is 10.4 keV ($\equiv 10.4 \cdot 10^{-11}$ rad cm^2/neutron). It was noted that the neutron flux was a factor of 2-3 less than predicted, the mean energy was higher by 46% and the gamma dose rate was a factor of 2 higher. It transpired that the thermal neutron capture cross-section of the argon component in the filter configuration had been incorrectly modelled[10]. Natural argon contains just 0.34% of ^{36}Ar, the rest being ^{40}Ar. The contribution of ^{36}Ar to the overall capture cross-section of natural argon had been ignored. However, the neutron capture cross-section for ^{36}Ar is approximately 100x greater than that for ^{40}Ar. Hence, effectively the overall cross-section of natural argon was underpredicted by about 1/3, which is in line with the reported measurements.

Revised calculations using the correct argon data now show good agreement between measured and calculated results. Hence, on the one hand confirming the lower than expected beam quality, but on the other hand, giving confidence to the calculations which are required to predict the irradiation doses in the programme of phantom, cell culture and healthy tissue tolerance experiments.

The measurements also confirmed that the beam is highly collimated with an angular spread of less than 10 degrees.

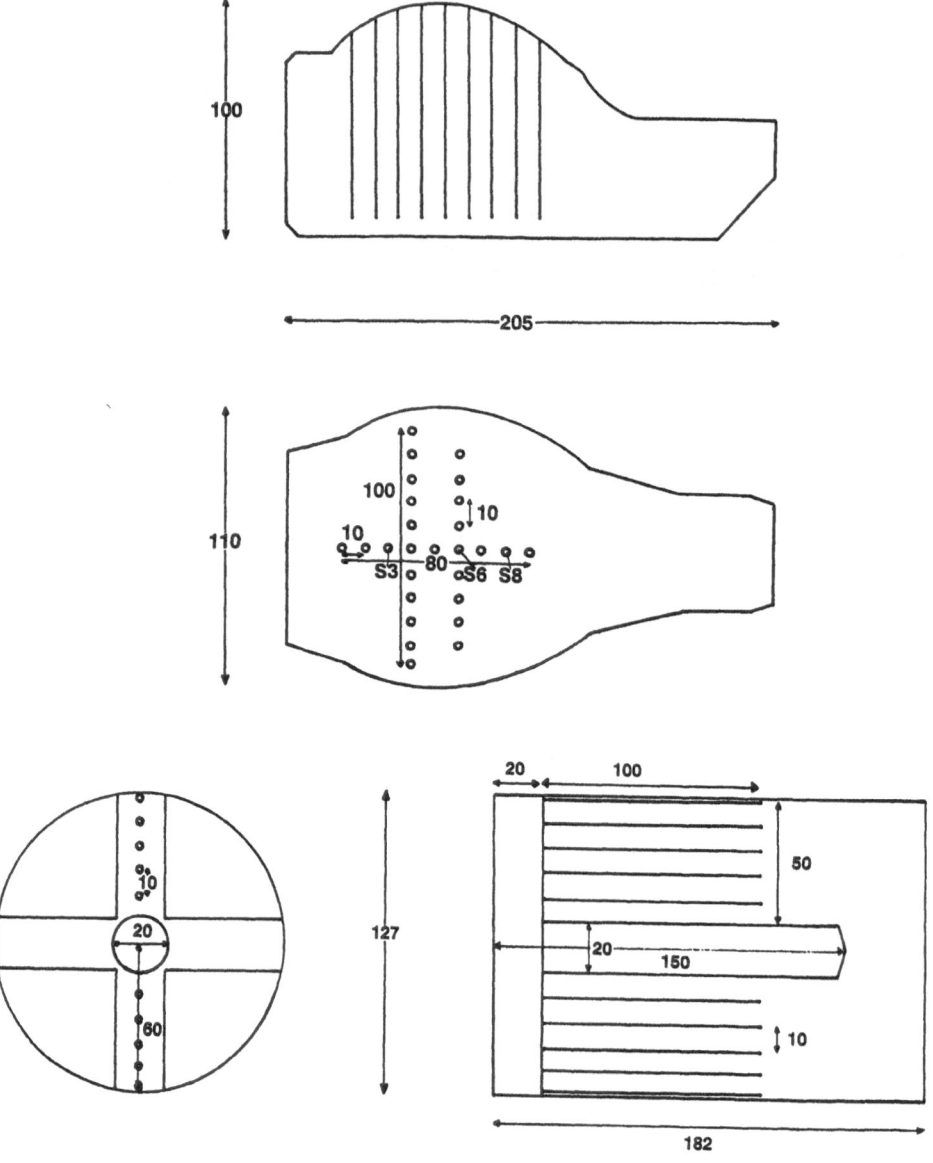

Figure 3. *The shapes and dimensions (in mm) of the beagle-head and cylindrical phantoms, showing the finely-drilled holes for placing foils and TLDs.*

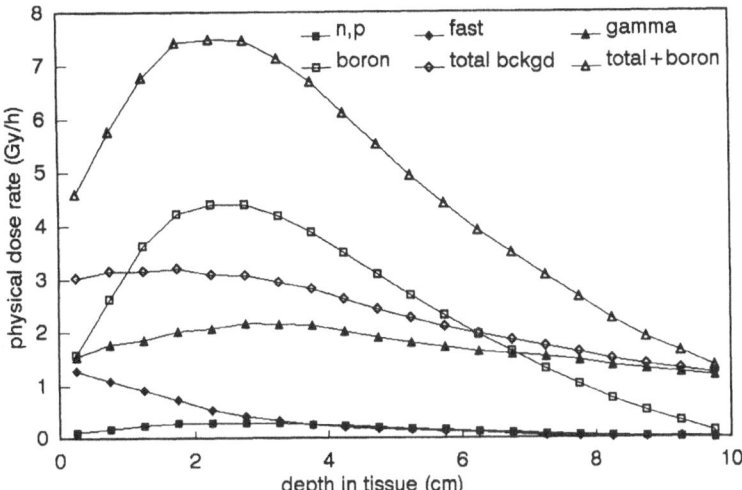

Figure 4. *Depth dose rate distributions in a tissue equivalent phantom for each beam component in the Petten epithermal neutron beam.*

To evaluate the neutron and gamma fields in plastic phantoms, two phantom experiments were performed. The first was a realistic model of a standard beagle head and the second, a solid cylindrical model. Both phantoms are constructed of "lucite" (acrylic plastic). The shapes and dimensions of the phantoms, which were provided by INEL, are shown in Fig. 3. Drilled holes distributed as indicated, allowed foils and TLDs to be placed. The comparisons between measurements and calculations are pending.

Cell Culture Experiments

The first cell culture experiments using the HB11 epithermal neutron beam were performed in July. V79 Chinese hamster cells in suspension were placed in small plastic phials at radially and axially distributed positions in a hollow, cylindrical, polyethylene phantom filled with tissue equivalent liquid. The experiments aim to verify and corroborate physical dosimetry and calculations of the beam in the presence of different concentrations of ^{10}B (initially with boric acid). For the dosimetry, activation foils and TLDs are placed on and within the phantom. The phantom itself is 16.6 cm in diameter and 23.0 cm in length. It is therefore of the same dimensions as that used for similar studies at the Brookhaven Medical Research Reactor (BMRR) epithermal beam and hence direct comparisons can be made. This programme continues, as at present insufficient data has been obtained to present conclusive results.

Table 1

*Programme of experiments for the healthy tissue
tolerance study on the canine brain*

Single fraction

mean blood ^{10}B conc. [ppm]	Peak physical dose to blood (2.25cm)[Gy]	number of dogs
25	15	5
	19	5
	23	5
50	23	5
	27	5
	31	5

Four fractions

25	15	5
	19	5
	23	5

Healthy Tissue Tolerance Studies

For the treatment of patients with malignant glioma, and from the healthy tissue tolerance studies in dogs performed at the BMRR, it is evident that the tissue at risk is not the skin but the normal brain. As depicted in Fig. 4, the peak physical dose for an epithermal beam occurs at 2-3 cm depth, i.e. in the healthy brain. Hence, as in conventional radiotherapy, the limiting dose here is the maximum dose to which the brain may be irradiated. It is therefore necessary to determine dose response curves for different mean blood boron (^{10}B) concentrations. The expected boron concentration during treatment will be probably 25-50 ppm ^{10}B in blood. The planned matrix of experiments[11] are shown in Table 1. The work has only recently started and will continue for another 6 months, followed by a 1-2 years period of observation. Both single and multi-fractionational experiments will be performed. The latter is intended to simulate as closely as possible the expected treatment scheme for the eventual patient irradiations[12].

From the experiments that began recently, 2 beagle dogs from the 23 Gy dose, single fraction, group were irradiated. After 4 weeks, both dogs had epilation at the irradiation field site. At present (2 months since the irradiation), both dogs have yet to show any neurological symptoms[11]. Both visual and magnetic resonance imaging observations are in progress. With the beam currently being overhauled, the healthy tissue tolerance study is currently in abeyance.

Treatment Planning

A treatment planning scheme, based on the code MCNP, is currently under development[13]. In its present form, the program can be used already to predict the irradiation doses for the phantom experiments and the healthy tissue tolerance studies. By means of spreadsheet analyses, the calculated peak doses at depth are derived based on the measured and expected mean blood-boron concentrations and the irradiation times, see for example Fig. 4. For BNCT, conventional treatment planning schemes which usually employ a semi-empirical approach, are of limited use only, as the neutrons in the incident beam, unlikely photons, will spread quite extensively in the target volume. Hence, the present code needs to be extended considerably by incorporating 3D treatment planning routines, with full 3D graphics capabilities. The code must be in a form that can be readily used by the radiotherapists for the pre-treatment assessments.

Treatment Room

The current set-up of the irradiation room, see Fig. 2, is only suitable for small scale experiments, such as those described above. The situation will suffice for the experimental programme over the coming 2 years. Thereafter, it will be necessary to enlarge the facility for patient irradiations. Nevertheless, in order to provide sufficient shielding of the irradiation room, even in its present, small configuration, over 40 tonnes of shielding in the form of heavy concrete, lead bricks and borated polyethylene sheets have been installed. Also, operation of the beam shutters, the liquid argon system, the dosimetry/monitoring instrumentation, is all performed semi-automatically at present. Extension to a full-scale, automated treatment facility will be no mean task, and will depend equally on the requirements specified by the radiotherapists at the Netherlands Cancer Institute and the nuclear physicists at Petten.

ON THE APPROACH OF CLINICAL TRIALS

The start of clinical trials are presently planned for the end of 1992 or beginning of 1993. Before the trials may start, it will certainly be necessary to complete the nuclear and biological characterisation of the beam. In addition, the supplementary activities currently in progress at other centres in Europe must also be concluded. These include: pharmacokinetic studies with BSH in human patients, which have started already with hospitals at

Bremen, Lausanne and Lund participating (see these proceedings); fractionated pharmacokinetic studies to investigate, after repeated administration of BSH, if a modified fractionation schedule is needed; a scheme for the provision of patient data on tumour size and location, to be incorporated into the treatment planning scheme; and the start of post-operative BNCT treatment of glioblastoma patients, as defined by and in agreement with, the procedures and protocols of the clinicians within the European Collaboration group[14]. Last, but not least, the irradiation room and the provisions for monitoring and observation at the HB11 facility must all be ready and operable.

Installation of the New Irradiation Room

The details of the design requirements for the irradiation room for the treatment of glioma patients have still to be finalised. At present, the maximum space available that can be used, is approximately 4 x 5 m² in floor space, with a maximum height of 2m. The working area or treatment area around the beam exit must be of sufficient size to allow the head of the patient, in lateral recumbence, to be irradiated from two laterally opposed directions. The minimum requirement is therefore a semi-circular area of 2 m radius, centred at the therapy position. The potential, available space at HB11 should therefore be sufficient, see Fig. 1.

Apart from creating space, emphasis on shielding requirements cannot be understated. Thorough shielding calculations, the quantity and size of shielding material will need to be determined. Within the room, provisions must be available to monitor the beam, position the patient, observe the patient, and place the patient for treatment.

Provision of Observation Area for Medical Staff

During treatment, a working area must be made available outside the shielded area of the treatment room for placing instrumentation for medical observation purposes. An instrumentation panel to display the following information will be required : the total neutron and gamma fluxes; the neutron and gamma dose rates; and audio/visual contact with the patient.

Facilities for Reception of Patients

Access from outside to the reactor building and hence, the facility, will be via the reactor building's emergency exit which fortuitously is adjacent to the HB11 entrance, giving direct access, without hindrance to normal reactor operations.

For clinical trials, no permanent facilities are foreseen. It is considered that temporary accommodation, i.e. in the form of portacabins will suffice. It

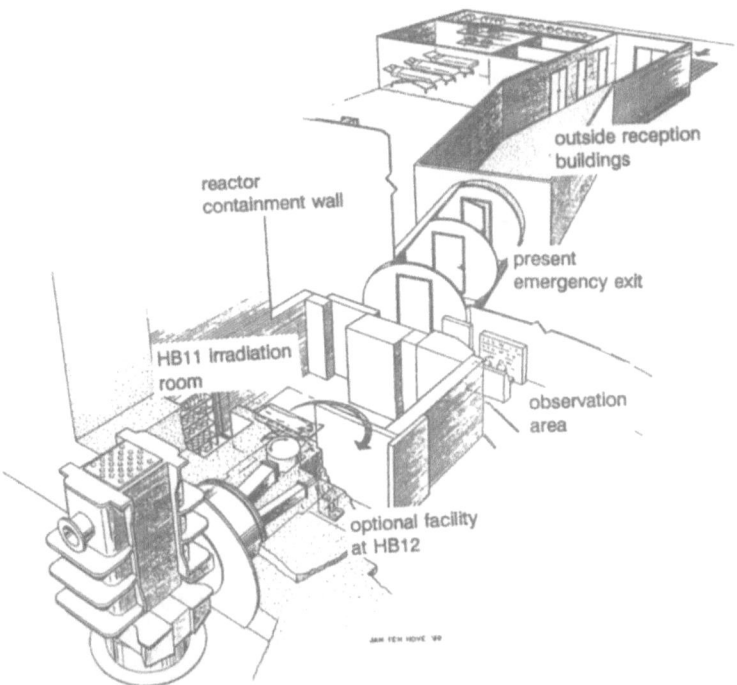

Figure 5. *Artist's impression of the treatment facility at HB11, indicating irradiation room, observation area, outside buildings, plus optional facility at HB12.*

is expected that patients will arrive by ambulance from the respective hospitals in The Netherlands (Amsterdam 55 km, Leiden 80 km). Following irradiation treatment, the patients will return directly to the hospital.

An artists impression showing an overview of the treatment room, beam tube, reactor vessel, observation area and outside reception buildings is shown in Fig. 5.

Operational Treatment Planning Scheme

The treatment planning scheme, as stipulated above, should be in a proven, operational form. The procedures to use the scheme to assess the irradiation schedule, i.e. number of fractions, multi-port irradiation fields, should be formulated for use in the conventional way.

Patient Selection, Hospitalisation and International Transport

It is anticipated that the first patients for clinical trials will invariably be from the University Hospital at Leiden. In conjunction with the Radiotherapy Department at the Cancer Institute in Amsterdam, and in consideration and agreement with the procedures and protocol defined by the European Collaboration group, the patients will be transported to Petten for treatment. Inclusion of patients from outside The Netherlands is at the discretion of the clinicians involved.

BEYOND CLINICAL TRIALS

Regular Treatment

The clinical trials phase will probably last up to 2 years and possibly longer, depending on the outcome of the initial trials. Once proven, it is planned that a more permanent structure should be provided for the reception of patients who could be accommodated in a small ward of up to 3-4 beds, plus other provisions, see Fig. 5. Regular treatment could involve up to 4-5 patient treatments per day, making it inefficient and inconvenient to have patients transported to and from Petten in teams of ambulances. In addition, with the on-set of regular treatment, improved observation and monitoring systems are planned, plus improvement to any (mechanical) operating system related to the facility itself. For example, it will probably be a legal requirement to guarantee that the treatment conditions are reproducible, i.e. on-line dosimetric equipment has to be developed and installed that must be reliable, durable and able to monitor all beam parameters. At present, most of these thoughts are merely projections of future needs and are still to be evolved in any detail.

Continued Research

With the on-set of clinical trials and regular treatment of glioma patients, the research activities at Petten will continue towards improving and optimising the treatment. Along the lines of research that will be followed, include: improvement of dosimetry techniques to monitor the beam parameters and to display the measurements via visual display units at the observation panel; to further improve the treatment planning scheme to accelerate treatment schemes associated with increased patient turnover; to test new boron compounds that become available, by means of cell culture and healthy tissue tolerance studies similar to the ones currently in progress; to improve the beam performance by re-design, considering for example, alternative filter materials combinations; and to develop at Petten, additional beam lines for BNCT, e.g. beam tube HB12 at the HFR, as shown in Fig. 5, and beam tube HN at the Low Flux Reactor at ECN, Petten.

CONCLUDING REMARKS

BNCT offers the opportunity to treat certain types of tumours that are presently inoperable or highly metastasised. With the development of better tumour seeking boron-containing compounds and the improvement in the quality and penetrability of neutron delivery systems, it is now feasible that BNCT can be successfully demonstrated at Petten within the next 2 years.

Before the trials can begin, certain pre-conditions must be determined, such as: the pharmacokinetics of BSH; the healthy tissue tolerance of the brain; and the predictability of dose distributions by means of treatment planning schemes.

The work at Petten is central to the programme of BNCT in Europe. With a successful demonstration of BNCT at Petten, it is anticipated that at other reactor and accelerator centres in Europe, other facilities will be developed for the treatment of tumours by BNCT.

REFERENCES

1. Moss, R.L.; Stecher-Rasmussen, F.; Ravensberg, K.; Constantine, G.; Watkins, P.R.D. Design, construction and installation of an epithermal neutron beam for BNCT at the high flux reactor Petten. In: *Progress in Neutron Capture Therapy for Cancer.* Allen, B.J.; Moore, D.E.; Harrington, B.V., eds., Plenum Press, New York 63-66 (1992).
2. Gabel, D. Approach to boron neutron capture therapy in Europe: goals of a European collaboration on boron neutron capture therapy. In: *Proc. of the 2nd Eur. Part. Acc. Conf., Nice.* Editions Frontières, Gif-sur-Yvette, 283-285 (1990).
3. Röttger, H.; Tas, A.; von der Hardt, P.; Voorbraak, W.P. High flux materials testing reactor Petten, characteristics of the facilities and standard irradiation devices. *EUR 5700 EN* (1986/87).
4. Moss, R.L. Progress towards boron neutron capture therapy at the high flux reactor Petten. In: *Neutron Beam Design, Development and Performance for Neutron Capture Therapy.* Harling, O.K.; Bernard, J.A.; Zamenhof, R.G., eds., Plenum Press, New York, 169-184 (1990).
5. Watkins, P.R.D.; Constantine, G.; Stecher-Rasmussen, F.; Freudenreich, W.; Moss, R.L.; Ricchena, R. MCNP calculations for the design and characterisation of the Petten BNCT epithermal neutron beam. In: *Progress in Neutron Capture Therapy for Cancer.* Allen, B.J.; Moore, D.E.; Harrington, B.V., eds., Plenum Press, New York, 71-77 (1992).
6. Briesmeister, J.F., ed. MCNP - a general Monte Carlo code for neutron and photon transport. *LA-7396-M, Rev.2.* (1986).
7. Perks, C.A.; Delafield, H.J. Neutron spectrometry measurements of the Petten HFR, HB11 neutron beam. In: *Boron Neutron Capture Therapy: Toward Clinical Trials of Glioma Treatment.* Gabel, D.; Moss, R.L., eds., Plenum Press, New York, 79 - 91 (1992).

8.   Harker, Y.D.; Amaro, C.R.; Anderl, R.A.; Watkins, P.R.D.; Voorbraak,
     W.P. Neutron and gamma measurements at the Petten High-Flux
     Reactor (HB-11). In: *Proc. of 1992 ANS Annual Meeting on Neutron
     Capture Therapy.* Boston, MA (1992).
9.   Stecher-Rasmussen, F.; Constantine, G.; Freudenreich, W.; de Haas, H.;
     Moss, R.L.; Paardekooper, A.; Ravensberg, K.; Verhagen, H.;
     Voorbraak, W.; Watkins, P.R.D. From filter installation to beam
     characterization. In: *Boron Neutron Capture Therapy: Toward Clini-
     cal Trials of Glioma Treatment.* Gabel, D.; Moss, R.L., eds., Plenum
     Press, New York, 59 - 77 (1992).
10. Watkins, P.; Konijnenberg, M.; Constantine, G.; Rief, H.; Ricchena, R.;
     de Haas, J.B.M.; Freudenreich, W. Review of the physics calculations
     performed for the BNCT facility at the HFR Petten. In: *Boron
     Neutron Capture Therapy: Toward Clinical Trials of Glioma Treat-
     ment.* Gabel, D.; Moss, R.L., eds., Plenum Press, New York, 47 - 58
     (1992).
11. Siefert, A.; Casado, J.; Moss, R.L.; Gavin, P.; Philipp, K.; Huiskamp, R.;
     Dühmke, E. Healthy tissue tolerance studies for BNCT at the high
     flux reactor Petten - first results. In: *Boron Neutron Capture Therapy:
     Toward Clinical Trials of Glioma Treatment.* Gabel, D.; Moss, R.L.,
     eds., Plenum Press, New York, 181 - 190 (1992).
12. Bartelink, H. A proposal for phase I clinical trials of glioma patients. In:
     *Boron Neutron Capture Therapy: Toward Clinical Trials of Glioma
     Treatment.* Gabel, D.; Moss, R.L., eds., Plenum Press, New York, 221
     - 223 (1992).
13. Watkins, P. Present status of the three-dimensional treatment planning
     methodologies for the Petten BNCT facility. In: *Boron Neutron Cap-
     ture Therapy: Toward Clinical Trials of Glioma Treatment.* Gabel, D.;
     Moss, R.L., eds., Plenum Press, New York, 101 - 109 (1992).
14. Fankhauser, H.; Stragliotto, G. Proposal for patient selection criteria
     and follow-up for BNCT in patients with supratentorial malignant
     gliomas. In: *Boron Neutron Capture Therapy: Toward Clinical Trials
     of Glioma Treatment.* Gabel, D.; Moss, R.L., eds., Plenum Press, New
     York, 213 - 220 (1992).

REVIEW OF THE PHYSICS CALCULATIONS PERFORMED

FOR THE BNCT FACILITY AT THE HFR PETTEN

Peter Watkins[a], Mark Konijnenberg[b],
Geoffrey Constantine[a], Herbert Rief[c],
R. Ricchena[c], Johan B.M. de Haas[d], and
Willi Freudenreich[d]

[a] Institute for Advanced Materials
Joint Research Centre Petten
NL-1755 ZG Petten, The Netherlands

[b] Netherlands Cancer Institute
NL-1066 CX Amsterdam, The Netherlands

[c] Institute for Safety Technology
Joint Research Centre Ispra
I-21020 Ispra, Italy

[d] The Netherlands Energy Research Foundation (ECN)
NL-1755 ZG Petten, The Netherlands

INTRODUCTION

The BNCT facility at the High Flux Reactor (HFR) Petten has been constructed within the existing HB11 beam tube. It consists of a combination of materials which filter the raw neutron and gamma fields emerging from the reactor removing the unwanted components, the fast and thermal neutrons plus the photons, without reducing the neutron intensity too severely. Such a device was installed in the HB11 beam tube in the summer of 1990. Prior to this event extensive calculations were undertaken to model the transport of neutrons and photons through potential filter configurations before a final optimized design was achieved. This paper provides a review of that process.

In addition, various other physics calculations were undertaken to estimate the shielding requirements and to support various experimental programmes. These items will also be addressed here.

Boron Neutron Capture Therapy, Edited by D. Gable
and R. Moss, Plenum Press, New York, 1992

47

HISTORICAL OVERVIEW

The BNCT project began at Petten in the late 1980's. The calculational work was begun by AEA Technology, Harwell because of their experience in this work[1-4]. In early 1989 initial calculations commenced for the HB7 beam tube which was planned as an animal experimental facility. However it soon became clear that the small HB7 beam was incapable of providing the required neutron intensity and attention turned to the larger HB11 facility. By the summer of 1990 the filter design had been completed and the filter installed at HB11[5]. During this same year the demise of the Materials Test Reactors at Harwell saw the BNCT effort in the UK drastically reduced. Fortunately the physics expertise was transferred to Petten where the work continued.

After installation of the filter in 1990 and until the middle of 1991 several low power experimental measurements were performed with the HB11 facility. The calculational programme continued, supporting the experimental measurements, refining the design calculations, and estimating the shielding requirements for full power operation. The majority of this work was performed at JRC Petten using Monte Carlo methods (MCNP). However parallel calculations were also performed at JRC Ispra using deterministic methods (DOT/DORT) and also at ECN Petten (DOT/MCNP).

The first full power characterization of the HB11 beam was performed in June 1991. Given the lengthy calculational campaign and minimal experimental validation it was not at all surprising that significant differences were observed between the calculations and measurements. The reasons for these are discussed later.

CALCULATIONAL METHODS

The BNCT facility at the HFR could only be installed within an existing beam tube, significant alterations to the basic structure of the reactor were not possible. This restriction implied that the therapy position would be a considerable distance (5 to 6 metres) from the reactor core. For the neutron physics design of the facility the situation was therefore a deep penetration problem necessitating transport theory methods. In addition the filter relies upon the "windows" and resonances in the cross-sections of its component materials for its effect. Therefore cross-section data was required in a continuous energy form or as a large number of energy groups to adequately represent this effect. Moreover both neutron and photon quantities were required so coupled (n,γ) calculations were deemed to be essential.

These points considerably restricted the computer codes suitable for the design calculations; either to 2 or 3 dimensional deterministic codes (DOT, DORT or TORT), or else to Monte Carlo methods (MORSE or MCNP). For the sort of geometries under consideration the deterministic approaches suffered from the need for expensive computing facilities and a lengthy preparation phase for the nuclear data.

In fact the Monte Carlo code MCNP[6] provided an almost ideal tool for much of the design work. It has an extensive continuous energy data library with minimal preprocessing. Moreover the ability of the Monte Carlo method to model complex geometries was a very definite asset. Furthermore MCNP contains methods which permitted many combinations of filter components to be examined in a single run. Although this particular approach relies upon an approximation it was considered to be adequate for the preliminary survey calculations.

INITIAL BEAM PARAMETERS

The target beam parameters for the HB11 facility were initially set as follows:

Neutron Intensity	>	$1.0E+9$ neutrons cm^{-2} s^{-1}
"Average Energy"	<	9.0 keV
Incident Photon Dose	<	60 cGy hr^{-1}

Subsequently the priorities changed so that the minimisation of the average energy was stressed at the expense of the neutron intensity. Reduction of the incident photon dose was viewed as a less important objective.

In an attempt to account for the biologically important fast neutrons a dose weighted "Average Energy" was employed. Thus if $D(E)$ is the ICRU-26 neutron dose function and Φ the neutron flux then:

$$\text{Average Dose} = \int_0^\infty D(E)\Phi(E)dE \bigg/ \int_0^\infty \Phi(E)dE$$

$$\text{Average Energy} = D^{-1} (\text{Average Dose})$$

THE MONTE CARLO MODEL

Version 3A of MCNP was used for all of the calculations reported here. Initially the standard MCNP library provided the cross-section data, although this excluded ENDF/B-V based data which is not available outside of the USA. Later in the study data based upon the JEF-1 evaluation became available and for some nuclides this was preferred.

Several geometrical models were employed of various sophistication; half and whole core representations, with and without control absorbers present. Most of the initial survey work used a simple half core model as shown in Fig. 1. Point detector tallies were used to record the neutron and photon flux data at the therapy position which varied from 5.3 to 5.8 m from the core. The exact position was determined rather later in the design phase after much of the shielding had been installed.

Although point detector tallies can produce spurious results in Monte Carlo simulations unless used with care, their use made it possible to use an approximate method of representing the filter components which permitted many configurations to be examined in a single Monte Carlo run. The point detector tally is such that at each event a "pseudo-particle" is created. The contribution of the "pseudo-particle" to the tally at the detector position is then given by an expression of the form

$$\Phi(E) \quad = \quad W \ p(E) \ \exp[-\int_{0}^{R} \Sigma_t(s,E)\,ds]$$

where W is the particle weight, p is the probability of scattering towards the tally position and R is the distance between the event and the tally position.

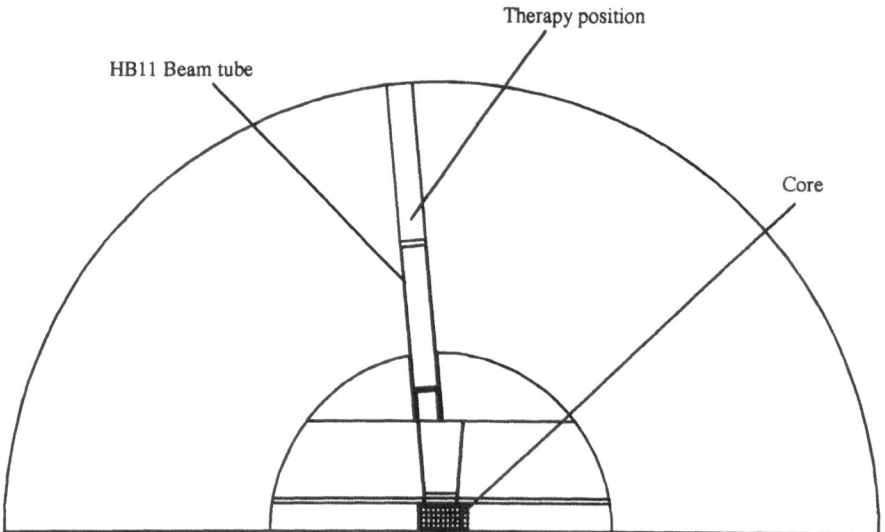

Figure 1. *Model geometry used for MNCP calculations.*

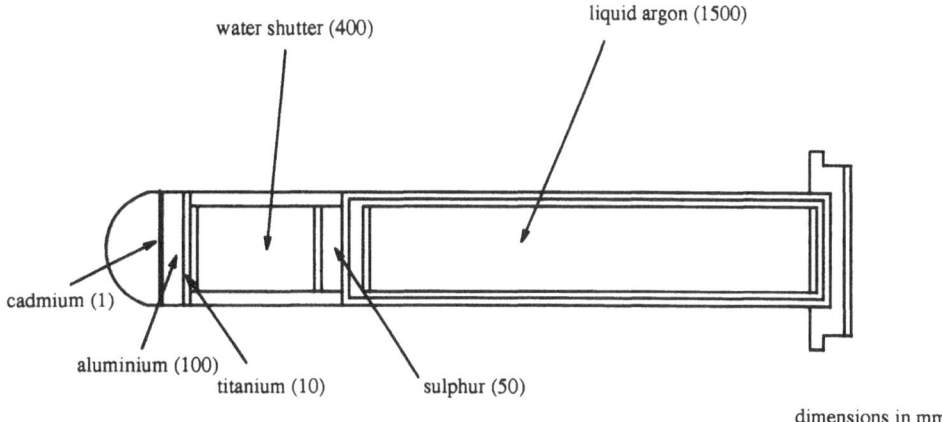

Figure 2. *As built filter configuration.*

MCNP permits any number of fictitious materials to be included between each event and the tally position simply by adding exponential terms to the above expression to model the attenuation through those materials. No additional particle tracking is required so the overhead on the Monte Carlo simulation is minimal and many different sets of such pseudo materials may be simulated in the same run. For the initial survey calculations this approach was used for all of the filter materials. Of course this technique is an approximation since neutron interactions within the filter materials are not represented, particularly multiple scattering and photon production. For the geometry under consideration the multiple scattering of neutrons within the filter is insignificant. The beam at HB11 is highly collimated simply because of the distance from the core so any neutron interaction within the filter effectively removes that neutron from the beam. The probability of a particle being scattered into the beam from within the filter is small and could be ignored. The photon production within the filter was recognised as a problem which would have to be addressed by more detailed modelling at a later date.

The use of argon in the filter caused considerable problems for the computer modelling because the standard nuclear data for this element was old and inaccurate. This problem was partially overcome by calculating energy dependent attenuation factors for argon from more recent data[7] and including these as point detector attenuators. This solution was not particularly elegant but was sufficient for the survey calculations. Later in the study better data became available and more realistic simulations were performed.

RESULTS FROM THE MONTE CARLO SIMULATIONS

Several hundred filter arrangements were examined with the components modelled as tally attenuators as discussed above. The results indicated that the "best" filter resulted from 15 cm aluminium, 5 cm sulphur, 1 cm titanium, 0.1 cm cadmium and 150.0 cm liquid argon. The calculated beam parameters for this arrangement are given in Table 1 as case P17(a) and are close to the target values.

Table 1

Results from selected MCNP simulations

Case	Filter					Therapy Position (m)	Neutron Intensity (n cm^{-2} s^{-1})	Average Energy (keV)	Photon Dose (cGy/hr)
	Al (cm)	S (cm)	Ti (cm)	Cd (cm)	Ar (cm)				
P17(a)	15	5	1	0.05	150.0	5.717	0.80E+9	7.4	37
P17(b)	10	5	1	0.05	150.0	5.4*	1.40E+9*	8.8	62*
Ispra	10	5	1	0.05	150.0	5.4	1.26E+9	6.1-7.8	60
S7X	15	5	1	0.1	150.0	5.76	1.46E+9	6.0	81
W3	15	5 .	1	0.1	150.0	5.76	0.69E+9	10.4	66

*Adjusted by inverse square to 5.4 m.

Note that the photon doses given here exclude photons generated from both argon activation and core fission products. The contribution of the photons from the fission products was estimated separately and found to give an extra 10 cGy hr^{-1}.

Construction of the filter began as soon as the material components had been determined. For engineering reasons the actual filter assembly was somewhat different than had originally been envisaged, as shown in Fig. 2. In particular a water shutter was included which could be flooded to reduce photon, and especially neutron, dose levels when the facility was not being used. Calculations were repeated with the real filter assembly modelled explicitly, except for the argon. The predicted beam parameters were close to those already estimated by the simple attenuation model. Flooding the water shutter was found to reduce the neutron intensity by 3 to 4 orders of magnitude and the photon intensity by a factor of 4.

Using a similar filter configuration to the optimal MCNP result, and with an input neutron source, JRC Ispra evaluated the beam parameters using the

DORT code in RZ geometry. Results for a therapy position 5.4 m from the core edge are shown in Table 1 as case Ispra. These calculations used cross section data based upon ENDF/B-VI and processed by NJOY-89. The agreement with the MCNP results was taken to be acceptable and were a good validation of the cross-section data.

As the study continued new MCNP cross-section library data became available which were based upon the JEF-1 evaluation[8]. In particular improved data for the main argon isotope, ^{40}Ar, permitted the argon chamber to be modelled explicitly. Unfortunately the ^{40}Ar evaluation is incomplete as there is no photon production data, but this appears to be the case for all other argon evaluations also. Using this data in a whole core model produced the results of case S7X in Table 1. Note the increased photon dose which was identified as being due to neutron scattering within the argon which was now modelled correctly and which gives rise to an additional capture gamma source within the support structure of the beam tube. The markedly higher neutron intensity was attributed as the effect of improved argon data. This assumption was later shown to be incorrect.

COMPARISON OF CALCULATION AND MEASUREMENT

The free beam parameters of HB11 at full power were first measured in June 1991 by INEL and ECN personnel using activation foil techniques[9]. In addition proton recoil measurements of the fast neutron components were undertaken at lower powers[10]. Combining all of the measured results and comparing with calculated estimates produced the following conclusions:

- Measured total neutron intensity - 4.9E+8 neutrons cm^{-2} s^{-1}. This was a factor 3 lower than the calculated estimate of 1.46E+9 neutrons cm^{-2} s^{-1}.
- Measured fast flux (> 1 keV) - 1.0E+8 neutrons cm^{-2} s^{-1}. Compared favourably with the calculated value of 9.3E+7 neutrons cm^{-2} s^{-1}.
- Measured photon dose - 105 to 120 cGy hr^{-1}. Some 50% higher than the calculated value.

Close examination of the data indicated that the calculations overpredicted the neutron flux by a factor of about 3 over the energy range up to 20 keV and underpredicted the flux by a factor of 5 above 1 MeV. However the fluxes above 1 MeV were sufficiently low that the effect on the fast neutron dose was small.

The principal reason for the large discrepancy between calculation and measurement was identified as being due to deficiencies in the argon data. Natural argon consists of 99.6% ^{40}Ar and 0.337% ^{36}Ar. It became clear after a careful examination of the problem that the ^{36}Ar cannot be neglected.

ARGON DATA

The ^{40}Ar data in the JEF-1 library appears to be a considerable improvement over the standard MCNP data. However the small quantity of ^{36}Ar present in natural argon has a dramatic effect on the macroscopic cross-section. The total cross-sections for both isotopes are shown in Fig. 3 where it can be seen that the ^{36}Ar has a cross-section some 100 times that of ^{40}Ar over the important epithermal region. The neutron attenuation through the liquid argon is thereby greatly enhanced by the presence of the ^{36}Ar. Comparing the results of cases W3 and S7X of Table 1, with and without the ^{36}Ar respectively, illustrates the effect dramatically.

The most recent model included a detailed representation of the shutters and all of the filter components. Natural argon, including the correct quantity of ^{36}Ar was included explicitly and the beam parameters at the actual therapy position (5.76m from core boundary) were estimated as:

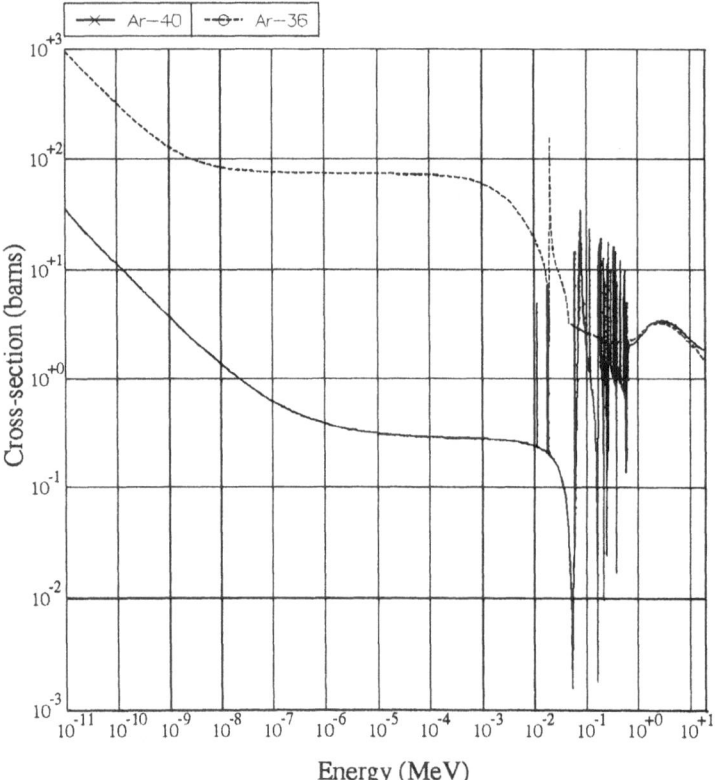

Figure 3. *Total cross-sections for ^{36}Ar and ^{40}Ar from the JEF-1 library.*

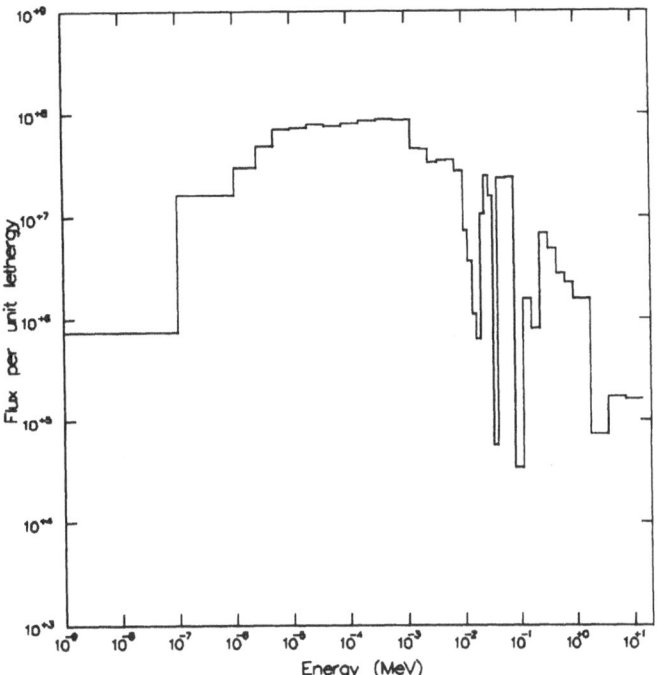

Figure 4. *Calculated Neutron Spectrum at the Therapy Position for HB11.*

Neutron Intensity	=	6.9E+8 neutrons cm^{-2} s^{-1}
Average Energy	=	10.4 keV
Photon Dose	=	66 cGy hr^{-1}

The neutron intensity in particular was now in much closer agreement with the measured values. There remains a discrepancy in the photon dose predictions but this can largely be explained if the photon production in the argon is included together with other minor photon sources such as the core fission products.

For the definitive geometrical configuration we have also performed calculations with the fluxes estimated by tallies other than point detectors. These have provided additional information but required very large, 10^7, numbers of particles to be tracked. From these the beam divergence at the therapy position has been estimated as less than 10° (half angle). This is an important parameter for input into the treatment planning systems which determine the dose distributions within the target and the irradiation time. It was also possible to derive a number of subsidiary particle sources at fixed positions along the beam tube. These could be used in subsequent calculations, for example the shielding work, to greatly improve the calculational efficiency.

The "best" neutron spectrum at the therapy position which has been derived from these calculations is shown in Fig. 4. Together with the neutron intensity this provides the basic input data for the treatment planning calculations upon which the irradiations depend.

SHIELDING CALCULATIONS

Prior to full power irradiations several surveys were made to determine the expected neutron and photon dose levels at various positions throughout the irradiation room. The results were then used to estimate the quantity and positioning of additional shielding material.

A detailed model of the irradiation room was constructed, Fig. 5, and MCNP calculations performed both with, and without, a water phantom at the therapy position. It was impossible to model the whole reactor core and the complete HB11 facility. Instead the calculations were divided into two. First a lengthy core calculation generated neutron and photon intensities and spectra at chosen planes within the HB11 beam tube. In this case the irradiation room was modelled in only minimal detail. Subsequent calculations ignored the core detail and much of the beam tube but included a detailed representation of the irradiation room. The source terms for the latter calculations being the already derived source planes within the beam tube. This two part scheme allowed us to make relatively crude estimates of the neutron and

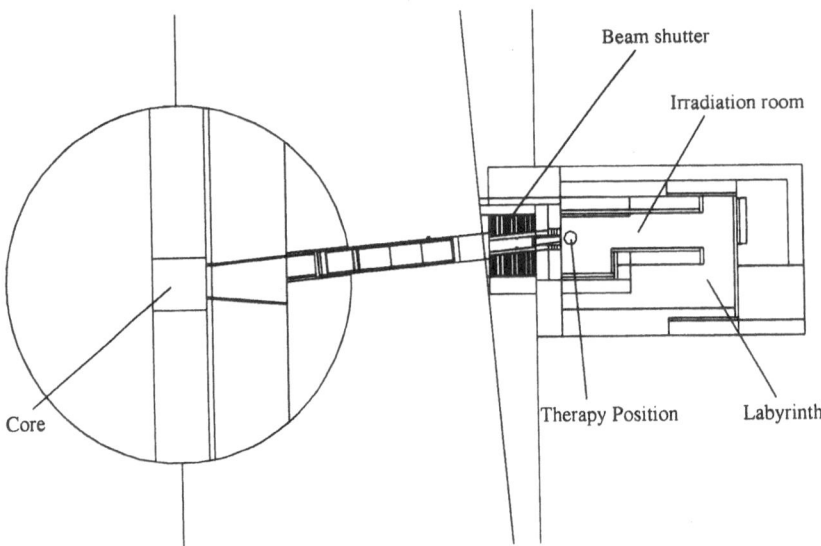

Figure 5. *MCNP Model of the Irradiation Room.*

photon dose rates at a large number of positions both inside and outside the irradiation room.

Although relatively crude these shielding calculations were of sufficient quality to indicate a number of potential shielding weaknesses, especially in the labyrinth and directly above the therapy position. The results were found to be in reasonable agreement with subsequent surveys of the irradiation room and its surrounds.

CONCLUSIONS

The methodology behind the design of the filter assembly for the Petten BNCT facility has been described. A considerable number of possible filter configurations were examined before an optimum was found. The calculated beam parameters resulting from the filter configuration have been compared with measured values and, initially, a considerable discrepancy was observed. To a large extent this was resolved by improving the cross-section data for one of the major filter components, namely the argon. It is believed that the remaining differences are also due to inadequacies in other cross-section data and not in the modelling because the results from several different researchers using various computer codes are in reasonable agreement.

Other applications of the calculational model have also been described. In particular the evaluation of the shielding requirements for the irradiation room produced very valuable data.

Future work will attempt to refine the model still further and eliminate the observed discrepancy between calculations and measurements. In addition alternative filter designs will be examined with the objective of improving the beam parameters for BNCT applications.

REFERENCES

1.　Constantine, G.; Baker, L.J.; Taylor, N.P. Improved methods for the generation of 24.5 keV neutron beams with possible application to boron neutron capture therapy. *Nucl. Instr. Meth.* A250, 565-572 (1986).
2.　Constantine, G.; Morgan, G.R.; Taylor, N.P. Progress towards boron neutron capture therapy at Harwell. *Int. Atomic Energy Agency, Vienna, 403-411 (1988).*
3.　Constantine, G.; Gibson, J.A.B.; Harrison, K.G.; Schofield, P. Harwell research on beams for neutron capture therapy. *Third International Symposium on Neutron Capture Therapy, Bremen, 30 May - 3 June, 1988.*
4.　Constantine, G. Neutron capture therapy beam design at Harwell. *Neutron Beam Design, Development, and Performance for Neutron*

Capture Therapy. Harling, O.K.; Bernard, J.A.; Zamenhof, R.G., eds, Plenum Press New York, 71-82 (1990).

5. Moss, R.L.; Stecher-Rasmussen, F.; Ravensberg, K.; Constantine, G.; Watkins, P. Design, construction and installation of an epithermal neutron beam for BNCT at the high flux reactor at Petten. In: *Progress in Neutron Capture Therapy for Cancer.* Allen, B.J.; Moore, D.E.; Harrington, B.V., eds., Plenum Press, New York, 63-66 (1992).

6. Briesmeister, J.F., ed. MCNP - a general Monte Carlo code for neutron and photon transport. *LA-7396-M, Revision 2, September 1986.*

7. Mughabghab, S.F. *Neutron Cross Sections* 1A, (1981), 1B (1984) and 2, BNL-325, Academic Press.

8. Vontobel, P. EFF1LIB: A NJOY generated neutron data library based on EFF-1 for the continuous energy Monte Carlo code MCNP. *Paul Scherrer Institute report no. 107, Sept. 1991.*

9. Harker, Y.D., Amaro, C.R; Watkins, P.R.D.; Voorbraak, W.P. Neutron and gamma measurements at the Petten High Flux Reactor (HB-11) beam. In: *Proc. of 1992 ANS Annual Meeting on Neutron Capture Therapy.* Boston, MA (1992).

10. Perks, C.A.; Delafield, H.J. Neutron energy spectrometry measurements of the Petten HFR, HB11 neutron beam. In: *Boron Neutron Capture Therapy: Toward Clinical Trials of Glioma Treatment.* Gabel, D.; Moss, R.L., eds., Plenum Press, New York, 79 - 91 (1992).

FROM FILTER INSTALLATION TO BEAM CHARACTERIZATION

Finn Stecher-Rasmussen[a], Geoff Constantine[b],
Willi Freudenreich[a], Han de Haas[a],
Raymond L. Moss[b], Ardi Paardekooper[a],
Klaas Ravensberg[a], Hans Verhagen[a],
Wim Voorbraak[a] and Peter R.D. Watkins[b]

[a] Netherlands Energy Research Foundation ECN
NL-1755 ZG Petten, The Netherlands

[b] Institute for Advanced Materials
JRC Petten
NL-1755 ZG Petten, The Netherlands

INTRODUCTION

The replacement of the vessel of the High Flux Reactor in Petten (1984) offered a unique opportunity to upgrade the neutron beam facilities. In the former thermal column, a Large Neutron Facility was installed, in which a target position views the entire surface of one side of the reactor core box. Through the increased solid angle a factor of about 10 higher neutron flux is obtained at the Large Neutron Facility as compared to the conventional beam holes of the HFR.

This large facility[1], HB11, formed the basis for the development of a neutron source suited for clinical BNCT. A neutron filter for transmission of epithermal neutrons was constructed and installed during the reactor's 1990 summer stop. The first series of experiments, as described in the present contribution, consists of a characterization of the free beam, dosimetry measurements of a cylindrical phantom and the dosimetry on the two first test animals as part of the healthy tissue tolerance study.

Boron Neutron Capture Therapy, Edited by D. Gable
and R. Moss, Plenum Press, New York, 1992

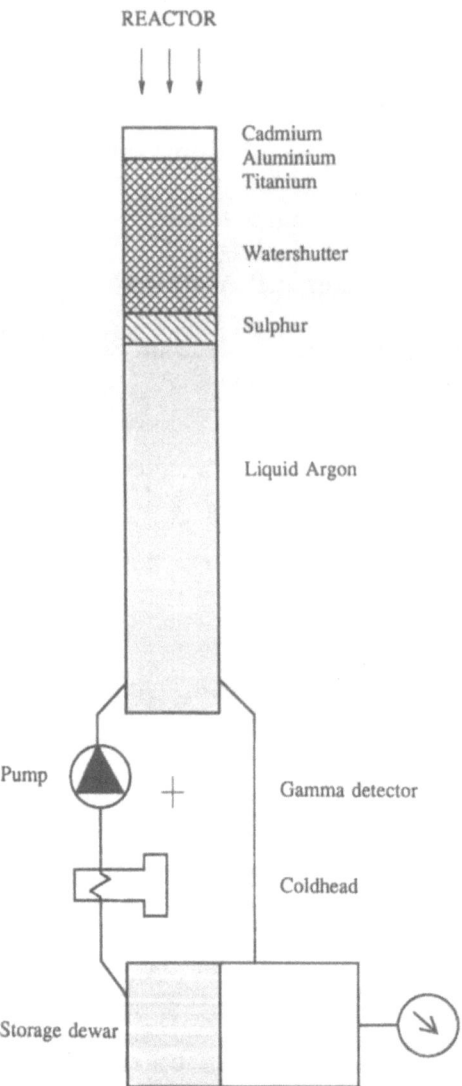

Figure 1. *Liquid argon filter.*

Table 1

Lengths of filter components

cadmium	aluminum	titanium	sulphur	argon (liquid)
0.10 cm	15.0 cm	1.0 cm	5.0 cm	150 cm

CONSTRUCTION OF NEUTRON BEAM FILTER

To determine the optimum composition of a neutron filter transmitting neutrons in the intermediate energy region (epithermal neutrons) with a minimum of gamma contamination detailed computer modelling of the reactor core and filter components was performed by the code MCNP[2] (Monte Carlo Neutrons and Photons). The following selection criteria were chosen:

i/ dose-weighted average neutron energy < 10 keV
ii/ neutron flux density $> 10^9$ cm^{-2} s^{-1}
iii/ gamma dose rate < 0.6 Gy h^{-1}.

The resulting filter composition is shown in Table 1 and sketched in Fig. 1.

A radial beam tube is inherently related to a large gamma component in the beam. Therefore appropriate suppression of the gammas is needed. As an optimal gamma filter with a low neutron attenuation liquid argon has been chosen. The application of a large volume (130 l) of liquid argon in a neutron beam tube, however, presents complex safety problems due to the neutron activation of the argon and the energy stored in the liquid gas. Much effort has

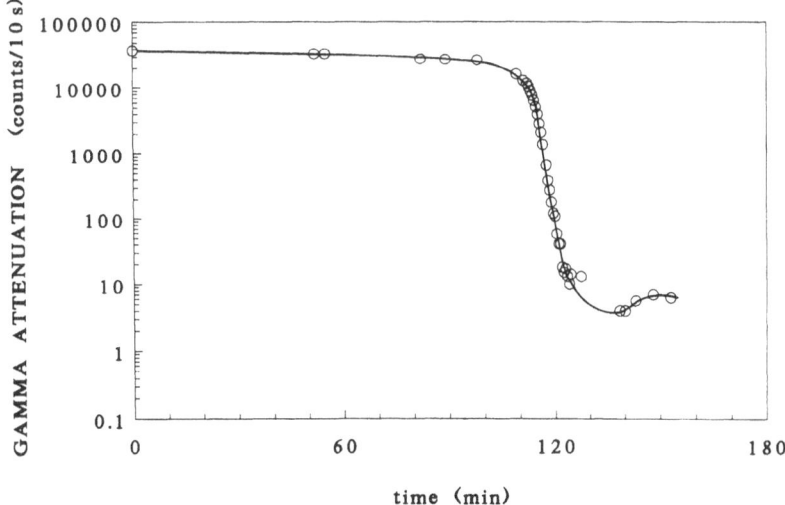

Figure 2. *Gamma transmission through argon filter. Filling the cryostat.*

been invested in safety measures for the argon system to fulfill the commissioning requirements.

Argon Filter System

Basically the argon system is very simple (Fig. 1): the liquid argon is circulated in a closed loop between a storage dewar and the beam cryostat. Before injecting the argon into the cryostat the liquid is cooled by a helium cold head, thus removing the nuclear heating of the argon. In this way the argon can be maintained in the liquid phase without evaporation and therefore without discharge of highly activated argon gas. However, the cryogenic design of the argon system is complicated by the fact that the liquid phase ranges from 84.4 Kelvin to 87.4 Kelvin at a pressure of 1 bar. Since the range of the liquid phase widens to 10 °C at a pressure of 2 bar it was considered safer to design the system for continuous operation at 2 bar.

For a reliable and safe technical realisation of the basic design additional features have been introduced: the cooling and pumping units are duplicated, a gas system has been connected to the cryogenic circuit and an emergency gas condensation system has been added.

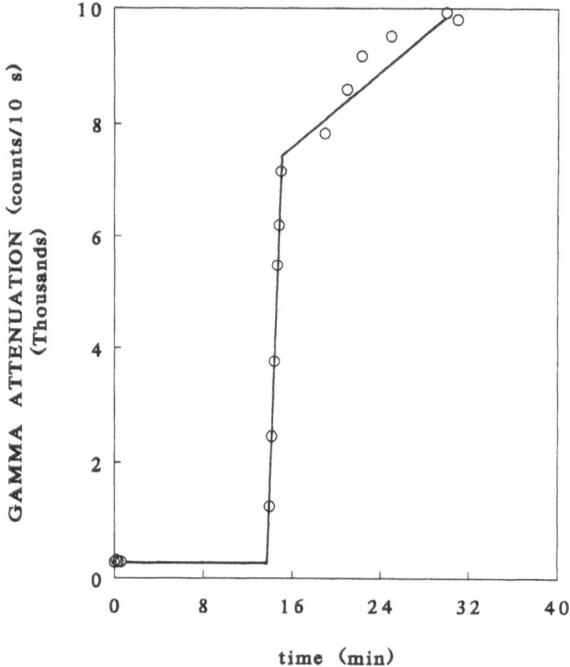

Figure 3. *Gamma transmission through argon filter. Effect of boiling.*

Table 2

Activation detectors for free beam characterization

Reaction	Composition of foil (weight %)		Dimensions (diameter x thickness) (mm)	Measured reaction rate and st. dev. (Bq/atom) (%)		Ratio meas/calc reaction rate
^{197}Au(n,γ)	1.0	Au	12.0x0.2	5.42E-14	3	0.374
^{115}In(n,γ) Cd*	0.2	In	12.0x0.1	6.46E-14	8	0.264
^{197}Au(n,γ) Cd	1.0	Au	12.0x0.2	5.04E-14	3	0.367
^{186}W(n,γ) Cd	1.0	W	12.0x0.2	1.98E-14	8	0.363
^{139}La(n,γ) Cd	5.0	La	12.0x0.1	3.80E-16	5	0.262
^{63}Cu(n,γ) Cd	10.0	Cu	12.0x0.2	1.76E-16	5	0.333
^{45}Sc(n,γ) Cd	99.99	Sc	12.0x0.2	3.26E-16	8	0.319
^{238}U(n,γ) Cd	22.83	^{238}U	12.0x0.2	9.01E-15	5	0.334
^{55}Mn(n,γ) Cd	1.0	Mn	12.0x0.2	3.94E-16	5	0.268
^{115}In(n,n') Cd	100	In	12.0x0.1	2.40E-19	15	4.31
^{58}Ni(n,p) Cd	100	Ni	12.0x0.2	< 8.5E-20		< 1.8

* irradiated with cadmium cover (see Fig. 5)

The performance of the argon system as a gamma filter is illustrated in Fig. 2. The curve shows the intensity of the gamma photons from the reactor core as measured with a GM-counter at the exit of the argon cryostat while filling the cryostat with liquid argon. To avoid interference from capture gammas, the measurement was performed during reactor shut-down. A gamma attenuation of a factor of $6.5 \cdot 10^3$ is desired from the measurement. It should be noted, however, that full cryogenic control is needed to safely construct and operate a liquid argon beam filter. To investigate the influence of external disturbance the pressure was abruptly decreased from 2.0 to 1.8 bar. This action lowered the corresponding boiling point resulting in vigorous boiling of the argon and increase of the gamma transmission, as is seen from Fig. 3. Therefore an on-line gamma-monitoring system will be essential as an ultimate check of the integrity of the argon filter.

FREE BEAM CHARACTERIZATION

Free beam characterization serves two purposes: i/ experimental verification of the MCNP calculations (to provide feed-back into the computer model) and ii/ actual determination of the beam parameters.

Gamma component in beam

As gamma detectors, thermoluminescent detectors (TLD) of the type 700 from Harshaw were used. The TLDs, encapsulated in cylindrical perspex holders (23.2/15 mm diameter, 7.9 mm height) for gamma build-up, were positioned along a horizontal and vertical line in the centre of the beam. Since the TLDs are enriched in ^{7}Li, and therefore insensitive to (thermal) neutrons, no correction for neutron capture in ^{6}Li was necessary for the determination of the gamma-component of the free beam having a negligible admixture of thermal neutrons.

A part of the free beam gammas originates from the decay of the activated argon ($t_{\frac{1}{2}}$ = 1.8 h). To investigate the influence of argon activation on the gamma component in the beam two separate gamma measurements were performed: 1) just after the start of the argon exposure and 2) after 5 hours of argon exposure. Extrapolating these measurements, summarized in Fig. 4, leads to the conclusion that the saturated argon activity gives rise to a 5% increase in beam gamma dose rate.

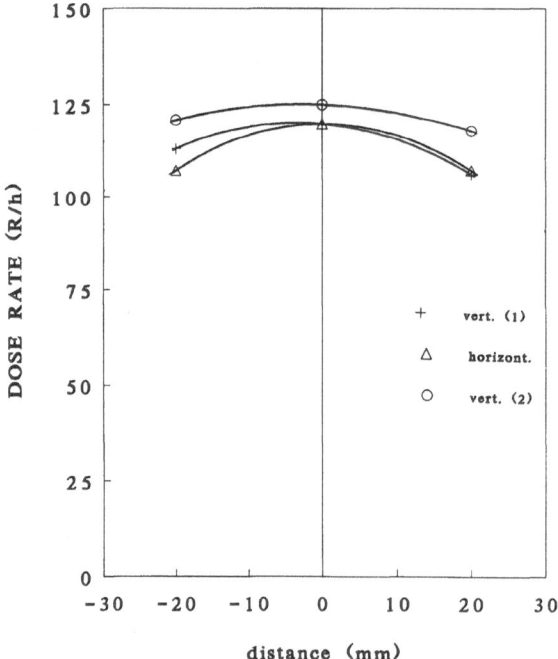

Figure 4. *Gamma profile of free beam.*

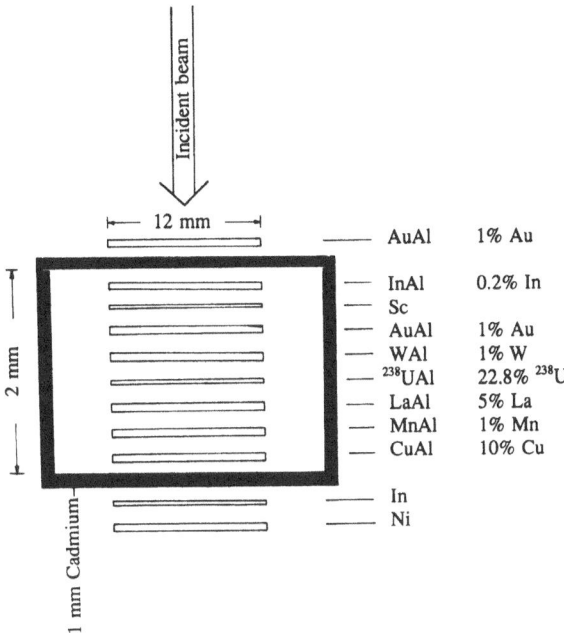

Figure 5. *Composition of monitor set for neutron spectrum measurement.*

Neutron Characterization

Several methods have been used by various laboratories to assess the neutron spectrum:

1. proton recoil spectrometers (Harwell)
2. Bonner spheres (Harwell)
3. activation detectors
    3.1. stacks of foils (INEL)
    3.2. thick foils (INEL)
    3.3. diluted foils (ECN)

Spectroscopy by methods 1. and 2. requires low counting rate and was therefore restricted to low power (5 to 50 kW) of the HFR. For the low power spectroscopy the assumption of the neutron energy distribution being independent of the power level of the reactor needs verification.

This paper describes the measurements by diluted foils as implemented by ECN. A spectrum monitor set consisting of a package of foils as sketched in Fig. 5 was irradiated at the therapy position. Note that the scale of Fig. 5 is exaggerated in the beam direction. The thicknesses of the foils are indicated in Table 2. The composition of the foils has been chosen to respond to neutron energies of the whole energy range of the epithermal neutron beam.

However, no suitable reactions within the important energy range from about 10 keV to almost 1 MeV were available.

Activation detectors measure reaction rates for the neutron activation reactions. After irradiation the gammas emitted by the activated foils were measured by four Ge-spectrometers coupled to a ND66 multi-channel analyzer. Due to the short available irradiation time the activities induced by the fast reactions were low. For foils of low activity special low-background counting equipment was used. Even so the uncertainty in the $^{116}In_m$ activity remained high (15%) while the activity from the Ni(n,p) reaction was just above the detection limit of 0.15 Bq. The Ni-reaction therefore was not included in the spectrum adjustment.

Activation detectors do not provide a direct measurement of the neutron energy distribution. Pre-knowledge of the neutron spectrum is needed to construct a spectrum from the measurements and thereby to derive a neutron flux. Since the concept of neutron flux is important for the description of neutron beams - especially for the sake of comparison of various beam facilities - an attempt was made to construct a spectrum. To this purpose the measured activation rates were compared with the calculated activation rates, obtained from a neutron spectrum in the irradiation position, computed by an MCNP-model of the reactor core and filter configuration (Fig. 6). The measured activation rates are presented in Table 2. The ratio between the measured and calculated activation rates, also in Table 2, clearly falls into two groups: a factor of about 0.3 for the reactions with low-energy neutrons and 4.3 for the only reaction with fast neutrons, $^{115}In(n,n')$. These ratios are then used to adjust the calculated neutron spectrum in such a way that the reaction rates, which are resulting from the adjusted spectrum, present a least-square fit to the measured reaction rates[3]. The resulting adjusted spectrum, which is shown in Fig. 6, is then finally used to generate the flux density: $\Phi = 4.2\cdot10^8$ cm^{-2} s^{-1}.

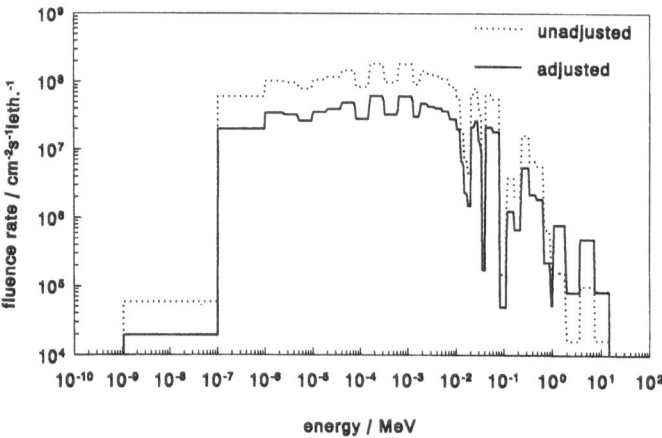

Figure 6. *Free beam neutron spectrum before and after spectrum adjustment.*

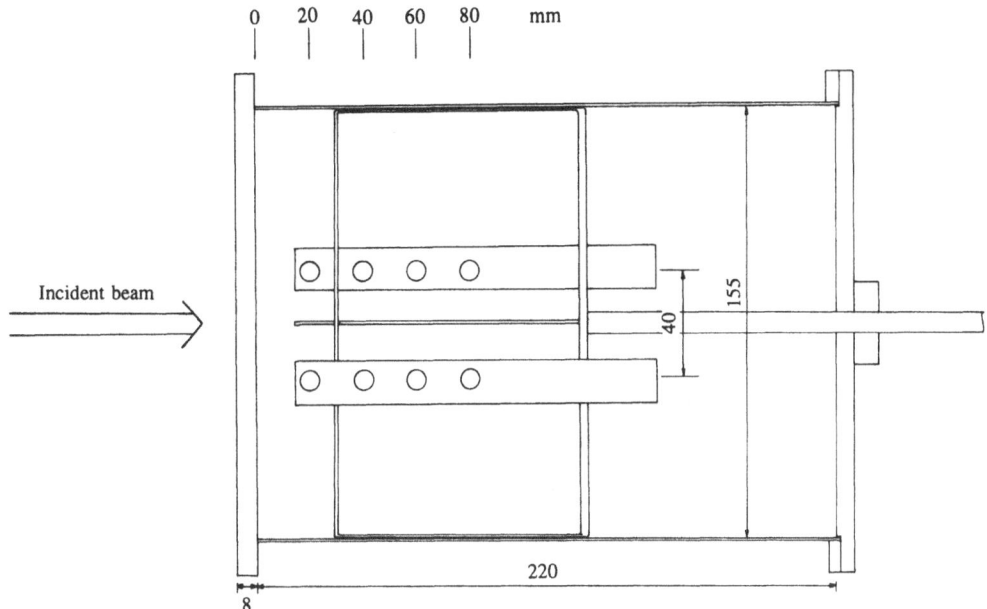

Figure 7. *Cylindrical phantom (University of Bremen).*

As the ratios between measured and calculated activation rates also serve the purpose of validating the computer model, on which the calculated values are based, a possible explanation for the observed deviation from unity could be found in the omission of the 0.33% abundant ^{36}Ar in the evaluation of the cross-section of argon.

TARGET DOSIMETRY

Experimental

The first series of experiments after characterization of the free beam included characterization of the beam in various phantoms and the first two canine (beagle) experiments to study the healthy tissue tolerance of brain tissue exposed to BNCT. Among the phantoms this report describes the dosimetry on a cylindrical, liquid filled phantom from the University of Bremen (Fig. 7), for which the dosimetry was performed by ECN.

Monitor sets consisting of an activation foil assembly and TLDs were used. Each activation foil assembly was comprised by a manganese-nickel foil (88% Mn), an aluminium-gold foil (1% Au) and a copper foil. The TLDs (enriched 7Li, HARSHAW type 700) were either encapsulated in a perspex cylindrical box (beagle "B") or in a small cylindrical polythene capsule (beagle "A" and the phantom). The reaction rates from the reactions, $^{197}Au(n,\gamma)^{198}Au$, $^{63}Cu(n,\gamma)^{64}Cu$ and $^{55}Mn(n,\gamma)^{56}Mn$, were determined by the same gamma-spectrometers as used for the monitor sets of the free beam characterization. The main resonance energies for the three reactions are at 4.9 eV, 337 keV and 580 keV, respectively. As the manganese and copper reactions have about the same response region, the spectrum index, i.e. the ratio between the reaction rates, will be almost independent of the neutron spectrum. This ratio presents a quality check of the measurements, demonstrated by a small variation coefficient.

The TLDs are slightly sensitive to thermal neutrons. In a moderating target, epithermal neutrons generate thermal neutrons which are captured by the small admixture of ^6Li in the TLDs and erroneously simulate an additional gamma field. Experimental determination of correction factors for the individual TLDs, actually used in the present work, is in progress. Until these experiments are completed, the best values from previous work is used[6]. This implies a correction for thermal neutrons of 1.09 cGy per 10^{10} neutrons cm^{-2}.

Derivation of Neutron Flux Densities from Reaction Rates

The energy spectrum of the free neutron beam (Fig. 6) contains hardly any neutrons of energy lower than 0.55 eV (cadmium cut-off energy). When a hydrogenous target is exposed to the free beam the epithermal neutrons are slowed down by the hydrogen nuclei. This process, which is responsible for the thermal flux peaking about 2.5 cm inside the target, changes the neutron spectrum according to the position. By approximation the spectrum at depth can be described as the spectrum of the free beam superimposed on a Maxwellian distribution, representing a neutron gas at thermal equilibrium. For the presentation of the flux densities it is convenient to divide the neutron spectrum into a thermal, Φ_{th}, and an epithermal part, Φ_{epi}. The sum of both parts gives the total flux density, Φ_{tot}.

$$\Phi_{th} = \int_0^{E_c} \Phi(E)dE; \qquad \Phi_{epi} = \int_{E_c}^{\infty} \Phi(E)dE$$

The value of the energy E_c plays an important role in comparing thermal and epithermal flux densities correctly. For E_c usually the cadmium cut-off energy (0.55 eV) is taken. However, to compare the flux densities of the present work with previous phantom characterizations at the Brookhaven Medical Research Reactor (BMRR)[4] a value of $E_c = 0.414$ eV has been adopted.

The reaction rate of a neutron activation monitor can be expressed as:

$$R = \sigma_{th} \cdot \Phi_{th} + \sigma_{epi} \cdot \Phi_{epi} ,$$

where σ_{th} and σ_{epi} are spectrum-weighted average cross-sections for thermal and epithermal neutrons. Again, for comparison between flux density distributions in various targets as measured in Petten and at the BMRR, the average cross-sections used for the BMRR dosimetry have been applied in the present work. It should be noted, however, that one must be careful in the interpretation of the absolute values of the flux densities in this simple two-group approach. To illustrate this, flux densities in a canine target have also been calculated without the simplifying approximation of the average cross-section, but by folding the energy dependent cross-section with the adjusted neutron spectrum in Petten. For this calculation, of which the results are presented in Table 5 between brackets, E_c has been taken equal to 0.425 eV.

Table 3

Thermal and epithermal neutron flux densities, physical neutron and gamma doses, and spectrum indices in cylindrical phantom

Position* (cm)	Neutron flux Φ_{th} (cm^{-2} s^{-1})	Neutron flux Φ_{epi} (cm^{-2} s^{-1})	Neutron dose rate (cGy/h)	Gamma dose rate (cGy/h)	Spectrum index Au/Mn	Mn/Cu
front	1.66E+8	3.60E+8	-	-	38.6	3.15
2.0	5.88E+8	3.11E+8	148	294	17.2	3.13
4.0	5.06E+8	1.34E+8	85	302	11.9	3.10
6.0	2.96E+8	4.00E+7	36	209	9.57	3.10
8.0	1.36E+8	1.20E+7	15	157	8.81	3.11

* See Fig. 7.

Conversion to Physical Dose

The physical neutron dose has been calculated using the dose functions of Bach and Caswell for wet tissue[5]. This dose function includes all neutron induced reactions: elastic and inelastic scattering, the (n,p), the (n,α) and the (n,γ) reactions with the elements constituting the wet tissue. The ^{10}B(n,α) reaction, however, is not included. All doses presented are calculated using the best estimates of the neutron spectra at the proper positions, and not calculated by the average cross-section approximation as in the case of the two-group flux densities. For the canine doses, the neutron spectra are obtained from MCNP calculations on a canine model adjusted by the experimental activation rates. Neutron spectra at two positions are shown in Fig. 10. For the cylindrical phantom no MCNP calculations were performed. Therefore, spectra were constructed by extending the low energy range of the calculated free beam spectrum with a Maxwellian distribution, and then at each individual monitor position adjusting the resulting spectrum with the experimental activation rates from that position.

As shown above, the comparison of measured with computer modelled activation rates for the free beam showed, apart from an overall over- estimate of the epithermal flux density, a considerable under-estimate of the fast neutron flux density. Therefore the neutron spectrum of the free beam, obtained by the adjusting the calculated spectrum with adjustment factors from the experimental activation rates, as determined from the free beam characterization (see Fig. 6), will have increased uncertainties for particular energy regions. This uncertainty in the neutron spectrum will propagate into the physical dose derived from the neutron spectrum and therefore depend on the position within the target: the more thermal neutrons present, the less propagation of the uncertainty in the high energy region of the neutron spectrum will occur. To investigate this point, the calculation of the dose has been subdivided into four energy regions, each one of a different uncertainty (Tables 4 and 6).

Table 4

*Physical neutron dose rates (in cGy/h) for different
neutron energy regions in cylindrical phantom*

| | Position in phantom (see fig. 7) | | | |
Energy region	20 mm	40 mm	60 mm	80 mm
$E < 1$ eV	47.1	44.6	24.9	11.5
1 eV $< E < 10$ keV	20.6	8.6	2.2	0.77
10 keV $< E < 500$ keV	60.0	24.9	6.9	1.7
$E > 500$ keV	20.6	6.9	2.2	0.77
Total (cGy/h)	148	85	36	15

Cylindrical Phantom

As an introduction to cell culture studies, the neutron and gamma doses in a cylindrical perspex phantom (Fig. 7), diameter 150 mm, length 220 mm, from the University of Bremen, were measured. The phantom, exposed end-on to the epithermal neutrons of HB 11, was filled with tissue equivalent liquid and provided with monitor sets (described above) at the front face and at four positions inside the phantom.

Table 5

Thermal and epithermal neutron flux densities,
physical neutron and gamma doses, and spectrum
indices for canine irradiations

Position[a]	Neutron flux Φ_{th} (cm^{-2} s^{-1})	Neutron flux Φ_{epi} (cm^{-2} s^{-1})	Total neutron dose rate (cGy/h)	Gamma dose rate (cGy/h)	Spectrum index Au/Mn	Mn/Cu
			Beagle "A"			
1	1.17E+8	3.27E+8	91	142	45.0	3.09
	(1.25E+8)	(4.41E+8)				
2	4.25E+7	5.36E+7	11	47	26.0	3.21
	(5.20E+7)	(5.70E+7)				
3	5.30E+7	6.19E+7	14	50	24.3	3.27
	(6.33E+7)	(7.00E+7)				
4	3.07E+7	3.01E+7	7	34	21.8	3.26
	(3.40E+7)	(3.71E+7)				
5	1.60E+6	5.74E+5		14	13.4	3.11
6	5.06E+7	1.62E+7		69	12.8	3.09
7				1.0[b]	9.10	
8				8.6[b]	11.7	
TFP[c]			92[c]			
			Beagle "B"			
1	9.11E+7	2.91E+8	77	154	48.8	3.10
	(1.05E+8)	(3.72E+8)				
2	8.10E+7	2.94E+8	79		53.2	3.07
	(9.29E+7)	(4.04E+8)		}17		
3	9.63E+7	3.00E+8	86		48.1	3.09
	(1.01E+8)	(4.41E+8)				
4	5.21E+7	2.76E+7	10	25	16.0	3.14
	(5.33E+7)	(3.67E+7)				
5	4.68E+6	1.24E+6		90	11.9	3.09
6	2.74E+7	1-20E+7		39	14.2	3.21
7				0.6[b]	8.80	
8				4.3[b]	9.95	
TFP[c]			92[c]			

[a] see text
[b] not corrected for thermal neutrons
[c] calculated value for position "thermal flux peaking" (2.75 cm inside head)

The thermal and fast flux densities are presented in Table 3 together with the neutron and gamma dose rates and the spectrum indices. Table 4 shows the contribution to the physical dose rate from the four energy regions discussed above. As seen from Fig. 8, summarizing the results, the reaction rate distributions show a maximum at 28 mm from the front side of the phantom. This is consistent with the calculated thermal flux peaking at 27.5 mm. The spectrum index gold/manganese decreases with increasing distance from the front face as expected from the fall in the contribution from epithermal neutrons with the depth. The spectrum index manganese/ copper, having a variation of 2.5%, serves as a quality check on the measurements.

The measured gamma dose has been corrected for the contribution from thermal neutrons as mentioned above. In Fig. 11, the present measurement is compared with the results from gamma dose measurements in a few other phantoms.

Canine Experiments (Healthy Tissue Tolerance Study)

The beagles "A" and "B", being the first two test animals in the healthy brain tissue tolerance study, were irradiated for a period of 105 minutes. The irradiations were interrupted for a few minutes for sampling the boron con-

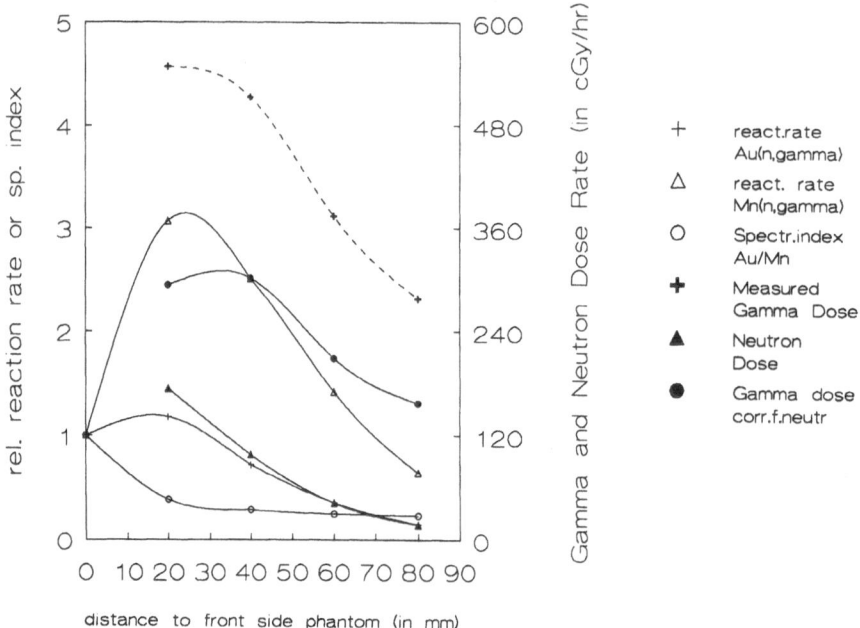

Figure 8. *Neutron and gamma measurements in cylindrical phantom.*

centration of the blood halfway through the session. Monitor sets were irradiated at the following positions:

1. top of the head
2. above the eyes, left
3. idem, right
4. rear of the skull
5. thyroid
6. in the mouth (fixed to the tubus)
7. chest (at the apex of the heart)
8. spinal cord (between shoulders).

The four most relevant positions are indicated on the sketch in Fig. 9.
Positioning of the monitor sets was done as accurately as possible, and the actual positions were recorded by photographs. However, as sticking plaster was
used to attach the monitors, the hair of the beagles hindered a close fixation
of the monitors on the skull. At positions 2 and 3 the monitor sets were located on the frontal bone in the supra orbital region (beagle "A") or between
the eyes (beagle "B"). The set for position 6 was attached to the respiration
tube, and therefore no photographic record of the actual position was made.

Three different types of encapsulation have been used for the TLDs: no encapsulation for the top of the head, a large perspex cylinder for beagle "B"
and a small polythene capsule for beagle "A". Further analysis will be needed
to determine the influence of the different encapsulations on the recorded
gamma dose. The measured gamma dose has been corrected for the contribution from thermal neutrons as mentioned above.

The physical dose rates have been calculated for the top of the head and
for the position of the thermal flux peaking (2.75 cm inside the head). To arrive at the neutron energy distributions at these positions the low energy part
of the spectra as calculated with MCNP was extended with a Maxwellian distribution. The resulting spectra were then normalized using the experimental
activation rates from the gold and manganese detectors for positions 1 to 4
and using the calculated gold and manganese reaction rates for the position of
the thermal flux peak. Fig. 10 shows examples of such spectra and a summary of the results is given in Table 5. The contribution to the physical dose
rate from each of the four energy regions discussed above is shown in Table 6
for the position at the top of the head (experimental) and at the thermal flux
peak (calculated).

A decrease of the reaction rate has been observed as the monitor positions
move away from the beam opening. The same holds for the gold/manganese
spectrum index, which gives a global indication of the ratio between thermal
and epithermal neutrons. For the free beam the index amounts to 128. The
index decreases to about 50 at the top of the head and is further reduced inside the tissue. It should be noted, however, that the value of this index for
position 1 (top of the head) is higher than the corresponding value for the
Bremen phantom (see Table 3 and 5). This is a clear indication of an unin-
tended separation between the monitor set and the surface of the skull,

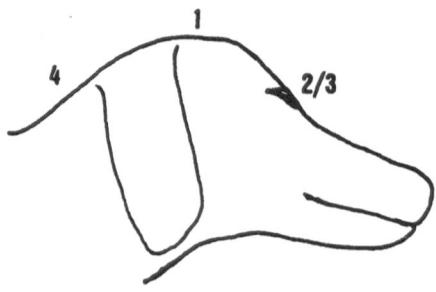

Neutron and Gamma Dose Beagle Bertha

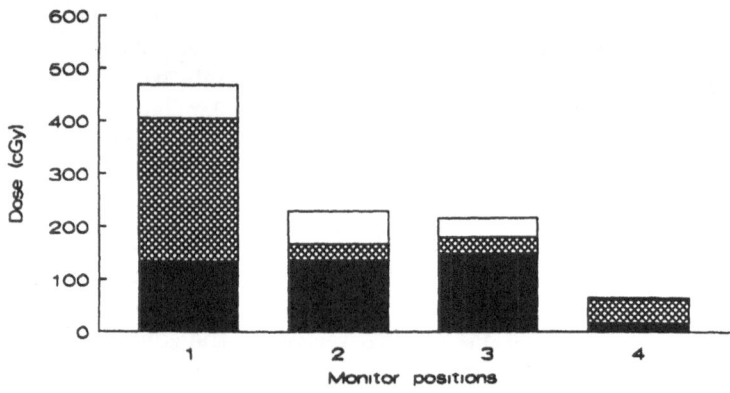

Neutron and Gamma Dose Beagle Anna

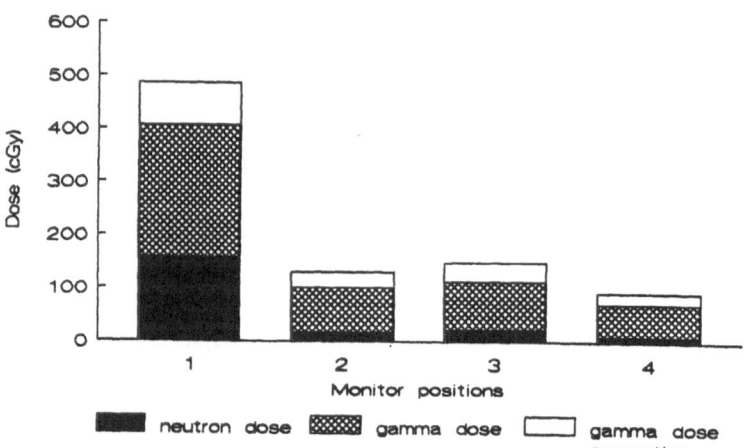

Figure 9. *Neutron and gamma doses from the canine irradiations.*

Table 6

Physical neutron dose rates (in cGy/h)
for different neutron energy regions in canine irradiations

	Neutron dose rate (cGy/h)		
	Measured at position 1 (top of the head)		Calculated for position "thermal flux peaking" (2.75 cm inside head)
Energy region	Beagle "A"	Beagle "B"	
E < 1 eV	12	10	58
1 eV < E < 10 keV	11	9	4
10 keV < E < 500 keV	54	46	21
E > 500 keV	14	12	9
Total (cGy/h)	91	77	92

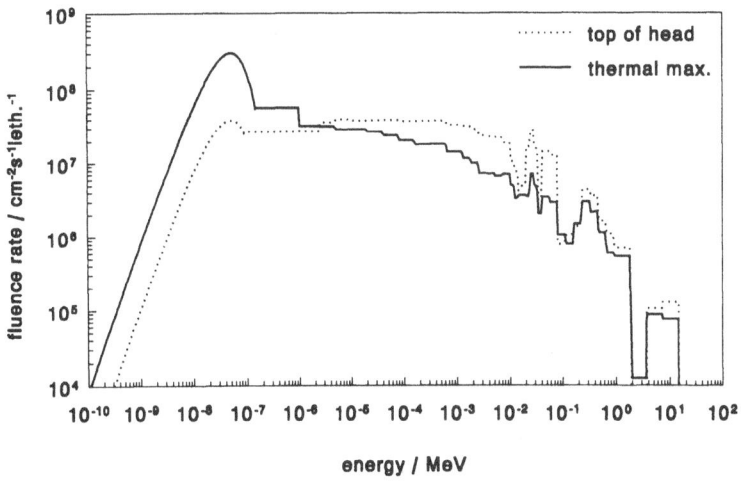

Figure 10. *Modified spectra for calculation of neutron doses.*

caused by the hair and loose skin of the dog. At the positions "chest" and "spinal cord" the quoted values are underestimated due to self-shielding of neutrons in the pure gold foils. The variation coefficient of the spectrum index manganese/copper (equal to 2.3%) indicates experimental consistency of the measurements.

Comparison of the neutron and gamma dose rates in Fig. 9, especially the monitor positions near the eyes, indicate a shift in the alignment of the dogs with respect to the beam aperture. For the future canine irradiations a more adequate alignment procedure will be developed.

DISCUSSION

Activation detectors do not measure neutron flux densities directly, but through an observable reaction rate, which is equal to the product of flux density and cross-section. As the cross-section is a function of the neutron energy, the flux densities derived from the measured neutron reaction rates will depend on the shape of the neutron energy spectrum. In a simple approach, where neutron spectra and cross-sections are presented as average values in a few energy groups (e.g. thermal and epithermal), the resulting flux densities strongly depend on the group representation of the spectra. Therefore the comparison of fluence rates requires a convention which should be adopted whenever flux densities from different BNCT facilities are compared. The

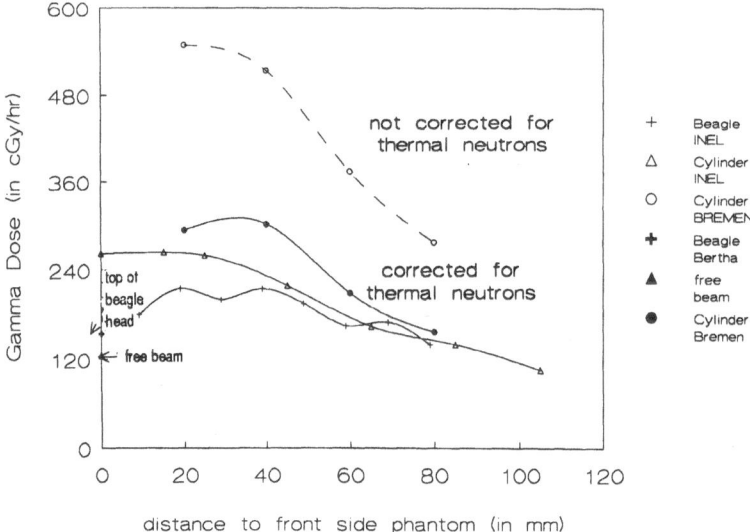

Figure 11. *Survey of gamma measurements in test animals and phantom.*

same considerations also apply to doses derived from activation measurements.

The discrepancy between the calculated and measured activation rates in the free beam can be solved by including the natural abundance (0.33%) of ^{36}Ar in the ^{40}Ar cross-section.

The experience gained from the present canine dosimetry will be employed in a protocol for the dosimetry of the remaining canine irradiations within the healthy tissue tolerance study.

REFERENCES

1. Stecher-Rasmussen, F. Upgrading the beam facilities of the High Flux Reactor in Petten. In: *Capture Gamma-Ray Spectroscopy and Related Topics - 1984. AIP Conference Proceedings number* 125, 933-935. Raman, S., ed., American Institute of Physics, New York (1985).
2. Watkins, P. Present status of the three dimensional treatment planning methodologies for the Petten BNCT facility. In: *Boron Neutron Capture Therapy: Toward Clinical Trials of Glioma Treatment.* Gabel, D.; Moss, R.L., eds., Plenum Press, New York 101 - 109 (1992).
3. Voorbraak, W.P. Freudenreich, W.E., Paardekooper, A., Stecher-Rasmussen, F., Verhagen, H.W. Neutron and gamma metrology in the free beam of HB11. *Report Netherlands Energy Research Foundation ECN-C--91-076, Petten November 1991.*
4. Harker, Y.D. Private communication.
5. Bach, R.L.; Caswell, R.S. Energy transfer to matter by neutrons. *Radiat. Res.* 35, 1-25 (1968).
6. Gibson, J.A.B. The relative tissue kerma sensitivity of thermoluminescent materials to neutrons. A review of available data. *Report EUR 10105.* Commission of the European Communities (1985).

NEUTRON SPECTROMETRY MEASUREMENTS OF THE

PETTEN HFR, HB11 NEUTRON BEAM

Christopher A. Perks and Howard J. Delafield

Radiation Dosimetry Department
AEA Environment and Energy
Harwell Laboratory
Oxfordshire, OX11 0RA, UK

INTRODUCTION

The Joint Research Centre (JRC), Petten, Netherlands, is currently at the forefront of boron neutron capture therapy (BNCT) research using intermediate energy neutrons. BNCT is a technique, under development, for the treatment of cancer, particularly glioblastoma, a fatal form of brain tumour. ^{10}B is preferentially introduced into the tumour cells and subsequent irradiation with thermal neutrons causes the ^{10}B to split, producing ^{7}Li ions and alpha particles. These both have short range (approximately equal to the diameter of the tumour cells) and are highly damaging to tissue. Hence, if the partition ratio of ^{10}B concentration between tumour and healthy cells is sufficiently large, the tumour cells are killed whilst the surrounding normal tissue is relatively undamaged. Because of the poor penetration of thermal neutrons, their use for BNCT applied to glioblastomas requires extensive surgery to reflect the scalp and open an aperture in the skull. Consequently, except in Japan, their use for BNCT is restricted to the development of treatments for melanomas. However, intermediate energy neutrons are able to penetrate surface tissue prior to thermalisation, thus avoiding the need for traumatic surgery.

Harwell Laboratory's involvement with dosimetry characterisation of filtered neutron beams for BNCT research dates back to the beams set up on the PLUTO and DIDO reactors at Harwell Laboratory for radiobiological experiments. The first beam constructed contained a filter of iron, aluminium and sulphur, which produced a high intensity beam of monoenergetic, 24 keV neutrons[1,2]. The second beam contained a filter of aluminium, sulphur and liquid argon. This produced a beam of neutrons with a broad energy range from 30 eV to 30 keV[3].

Following the Third International Symposium on Neutron Capture Therapy, Bremen, 1988, a 'Concerted Action' on BNCT, sponsored by the Commission of the European Communities (CEC), was initiated at Petten. This coordinates the various activities concerning BNCT in Western Europe. The first priority of this concerted action is to establish a BNCT facility at Petten with the aim of implementing BNCT for glioblastomas at the earliest possible date. An overview of the developments at Petten has been given by Moss[4]. Initial, trial neutron energy spectrometry measurements, were made using the HB7 beam of the High Flux Reactor (HFR), Petten, to confirm the accuracy of the computational techniques being used to design the therapy beam[5,6]. The HB11 beam is now being set up as an intermediate energy neutron beam. Neutrons emerging from the reactor core are filtered with a combination of aluminium, liquid argon, sulphur, titanium and cadmium[7]. First measurements of the neutron spectrum of this beam were made in October 1990. Unfortunately, after the measurements had been made, it was found that the argon filter was not completely filled. Consequently, the results from this measurement campaign have been discarded. A second measurement campaign was planned for June 1991 - this one was successful and the results are presented in this report.

The apparatus used in the spectrometry measurements is first described, followed by the methods adopted and a summary of the results. A full report on these measurements, and some subsidiary experiments, including measurements of the beam profile, is in preparation[8].

NEUTRON SPECTROMETER

A transportable high-resolution spectrometry system based on high-sensitivity cylindrical proton-recoil counters[9,10] and an alpha-recoil counter[11] has been developed at Harwell for measuring neutron spectra. The system was developed to overcome the limitations of the multisphere spectrometer, with its poor energy resolution and inability, in certain circumstances, to provide a unique spectral solution. For field measurements at higher dose rates, the cylindrical proton-recoil counters can be replaced by smaller volume spherical counters (type SP2) as used, for example, to measure the neutron spectrum inside the containment building of a PWR[12] and the previous measurements for BNCT research[1,3]. In the present application, measurements were made with the spherical proton-recoil counters (SP2) and the alpha-recoil counter.

Spherical Proton-Recoil Counters (Type SP2)

The neutron energy range (dependent upon the relative gamma-ray dose rate) which can be covered by the spherical counters is about 10 keV to 1.5

MeV. A set of three spherical counters was used, filled to nominal pressures of 1, 3 and 10 atm (1 atm = 101.325 kPa at 0°C) hydrogen gas. Their characteristics, operating conditions and calibration at the NPL have been reported by Birch, Peaple and Delafield[13]. Traceability of the calibrations is achieved for each counter by having a small internal alpha-emitting source which provides a pulse-height distribution of fixed energy in the counter, independent of the gas gain or electronic amplification. In addition, a trace of ^{3}He gas is included in the 3 and 10 atm counters providing an additional reference energy deposition of about 764 keV from the thermal neutron ^{3}He(n,p)T reaction.

The procedures for unfolding the measured pulse height spectra obtained during a field measurement with the counters to give the neutron spectrum are given by Birch, Marshall and Peaple[14]. A spline fitting procedure to the data is used, followed by unfolding with the SPEC4H program or an iterative program SCOFH.

Alpha-Recoil Counter

The alpha-recoil proportional counter[11] was developed to extend measurements above the upper energy limit (≈ 2 MeV) of the proton-recoil counters. The cylindrical counter, which has an active length of 300 mm and radius of 25 mm is filled with 6 atm ^{4}He plus 4 atm Ar to increase the stopping power. The counter has been calibrated at the NPL with monoenergetic neutrons at energies of 2.5, 3.75, 5.0 and 14.64 MeV. Subsequent measurements can be related back to the NPL calibrations by means of a ^{3}He reference peak. A reference edge in the pulse height distribution is also present at 5.3 MeV due to radioactive contaminants in the lead lining and solder components of the counter.

Good agreement was obtained between the measured response functions and those calculated by the HELIUM85 code, developed to generate the response function matrix[11]. Unfolding codes, used to obtain the neutron spectrum from the measured pulse-height distribution, have been developed. An iterative technique using the differentiated pulse-height distribution and a differentiated response function matrix was found to give the best results and has been adopted. Unfolding the measured response functions obtained at NPL, gave neutron fluences within $\pm 8\%$ of the values measured with a Precision Long Counter. A number of measurements were undertaken to determine whether the counter and unfolding programs operated satisfactorily for broad-energy neutron spectra. Measurements included the neutron spectrum from an ^{241}Am-Be source, and the leakage spectrum through a reactor shielding facility. The measured spectra showed very good agreement with the other measurements and calculations[11].

Data Acquisition

Electronics with simultaneous data acquisition are generally used. The pulses from each counter are fed via head amplifiers (EG&G Ortec type 142PC) into main amplifiers (EG&G Ortec type 572). Bi-polar pulses from the two head amplifiers are usually fed via a mixer-router (Canberra type 8222A) into memory segments of a multichannel analyser (Canberra series 35 Plus). The main amplifiers are used with 10 μs time constants for pulse shaping to ensure complete charge collection of the slowly rising pulses. Consequently count-rates are limited to below 1000 s^{-1} to avoid pulse pile-up. To optimise the low energy threshold of the measurements, the 1 atm proton-recoil counter was operated using separate electronics from the other three counters which were operated using the simultaneous data acquisition normally adopted. This technique enabled a low level discriminator setting to be used for the 1 atm counter, extending the spectrum measurements to the lowest neutron energy practicable (above the gamma-ray edge). The measured pulse-height-spectra were stored on floppy disc using a portable PC for subsequent analysis.

EXPERIMENTAL

As described above, the count-rates on the proton and alpha-recoil counters need to be less than 1000 s^{-1} to avoid pulse pile-up. Therefore, it was necessary to perform these measurements at very low reactor power and hence control the reactor power manually. Since it was anticipated that there would be fluctuations in the power level throughout the measurements, a BF_3 counter was positioned at the far end of the area to provide a monitor throughout the programme. The positions of the counters are shown in Fig. 1. At the end of the programme the Harwell BF_3 monitor was cross-calibrated against a BF_3 counter installed in the shielding surrounding the filter.

Neutron spectrometry measurements were made as near to the proposed dog irradiation position[15] as possible. To obtain the required data in the available time it was necessary to locate the counters one behind the other. Since the largest contribution to the neutron fluence and kerma-rates was expected from the lowest energy neutrons, the 1 atm counter was placed centrally in the beam with the higher pressure counters directly behind as shown in Fig. 1. The sequence of events during the experimental programme is given in Table 1. The first run at a nominal reactor power of 3 kW was done to ensure that the count-rate on the 1 atm counter was sufficiently low to avoid pulse pile-up. During this run the cadmium covers, used to minimise the incident thermal neutron fluence-rate, and, therefore the ^{3}He peak, were inadvertently left off the 3 and 10 atm counters. This error was found to be unimportant due to the relatively low thermal neutron fluence-rate in the incident beam. The second run at an increased power level (nominally 6 kW) was done to optimise the count-rates on the other three counters, which now all had their cadmium covers installed. Subsequent to the main runs (1 and 2), a number of short (≈ 200 s) runs (numbered 3 - 7) were made to assess the effect of the self shielding of the counters and the reduction in intensity with distance from the source. For all the runs, BF_3 monitor counts were recorded simultaneously. In addition, for the second run, ECN, Petten, staff simultaneously undertook activation foil measurements to determine the reactor power level.

The special nature of these measurements made it necessary to adopt a novel method of analysis. For the ^{4}He counter, since the relative neutron fluence in the energy range (> 2 MeV) of this counter was so small, it was not possible to unfold the neutron spectrum from the measured pulse-height distribution in the conventional way. The pulse-height distribution obtained during the first run was normalised to the power level of the second run (by comparison of the 1 atm proton-recoil counter count-rates) and then added to that obtained in Run 2. The counts in the channels corresponding to the

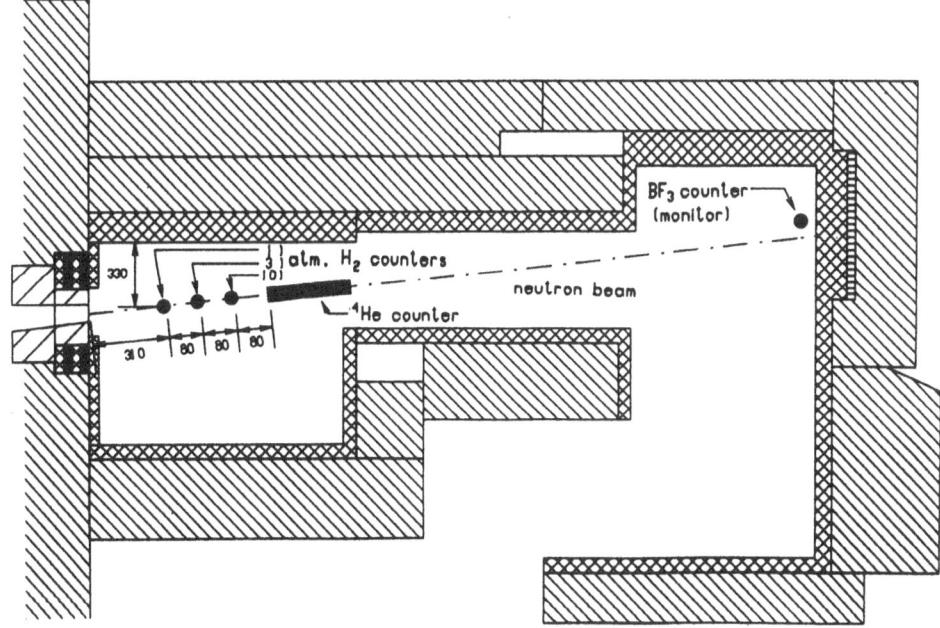

Figure 1. *Schematic diagram showing the positions of the proton and alpha-recoil counters and the BF_3 counter used as a monitor.*

energy ranges 1.5—2.5 MeV, 2.5—3.75 MeV, 3.75—5.0 MeV, 5.0—14.7 MeV and greater than 14.7 MeV were then summed and the corresponding background subtracted. The pulse-height distributions obtained for monoenergetic neutrons (at energies of 2.5, 3.75, 5.0 and 14.7 MeV) during the calibration of the counters were then used to estimate the neutron fluences in these energy ranges.

Table 1

Sequence of events during neutron spectrum measurements of HB11 beam, June 1991. (Power levels are nominal powers as requested during the measurements; later calibration gave the nominal power of 6 kW equal to 8.2 kW[7]

Time	Action
Before 06.00	Loading Argon filter
06.05	Reactor power raised to ≈ 3 kW
06.25 - 08.00	First beam profile measurement (172 mm from the beam opening)
08.05	Reactor power raised to ≈ 4 kW
08.12 - 09.18	Recoil counter neutron spectrometry measurement (Run 1)
09.45	Reactor power raised to ≈ 6 kW
09.54 - 12.26	Second recoil counter neutron spectrometry measurement simultaneously with activation foil measurements (Run 2)
13.00 - 13.40	Short runs with recoil counters to study self shielding effect and distance from source (Runs 3 - 7)
13.50 - 15.10	Second beam profile measurement (310 mm from the beam opening)
15.15	Comparison of BF_3 monitor with Petten BF_3 monitor

For the proton-recoil counters, the measured pulse-height distributions for both runs were normalised and added, channel by channel. The resulting pulse-height distributions were unfolded separately using the iterative unfolding method SCOFH. First the 10 atm counter was unfolded. The derived neutron fluence spectrum was then normalised to account for the reduction in intensity of the beam with distance to the 3 atm counter. The spectrum was then used to correct the measured pulse-height distribution of the 3 atm counter for its response to high energy neutrons before unfolding. Finally the process was repeated for the 1 atm counter. The neutron kerma-rate in water was calculated from the fluence-rate spectrum using fluence to kerma conversion factors given in ICRU[16].

Table 2

*Variation in count-rates of the proton and alpha recoil counters with
position in the beam*

(a) Position of counters

Run No.	Position I	Position II	Position III	Position IV
1	1 atm	3 atm	10 atm	4He[1]
2	1 atm	3 atm[1]	10 atm[1]	4He[1]
3	-	3 atm[1]	10 atm[1]	4He[1]
4	-	-	10 atm	4He[1]
5	-	-	1 atm	4He[1]
6	-	1 atm	-	4He[1]
7	1 atm	-	-	4He[1]

(b) Corresponding count-rates[2]

Run No.	Position I	Position II	Position III $\times 10^{-3}$	Position IV $\times 10^{-3}$
1	0.255 ± 0.001	0.199 ± 0.001	2.57 ± 0.08	4.26 ± 0.11
2	0.284 ± 0.001	0.221 ± 0.001	2.70 ± 0.08	4.26 ± 0.11
3		0.220 ± 0.001	2.43 ± 0.25	4.26 ± 0.11
4			2.73 ± 0.25	4.54 ± 0.32
5			0.208 ± 0.002	4.15 ± 0.30
6		0.216 ± 0.002		-
7	0.233 ± 0.001			-

Notes:

(1) Counter covered with 0.5 mm cadmium sleeve.
(2) Values given are count-rates summed in the following ranges:

1 atm counter	-	sum over channels 51 - 350
3 atm counter	-	sum over channels 50 - 300
10 atm counter	-	sum over channels 201 - 501
4He counter	-	sum over channels 61 - 511

corrected for dead time and normalised to unit count-rate on the BF_3 monitor.
Uncertainties are 1 standard deviation counting statistics.

RESULTS

The positions of the alpha and proton-recoil counters and the count-rates (corrected for dead time and normalised to unit count-rate on the BF_3 monitor) are given in Tables 2a and 2b. The count-rates are integrated over a range of channels for each counter taken to avoid the gamma-ray edge (for the proton-recoil counters), alpha edge for the alpha recoil counter, and the ^{3}He peaks. In the first run the cadmium covers were erroneously left off the 3 and 10 atm counters. This omission significantly affected the count-rate of the BF_3 monitor. However, if the results for Run 1 are normalised to the count-rate on the 1 atm counter during Run 2, the 3 and 10 atm counter and the ^{4}He counter all gave the same normalised count-rate during Run 1, within the uncertainties. The results from Runs 2, 3 and 4 show that any corrections due to the self shielding of the counters were negligible and hidden within the uncertainties of these measurements. The results for Runs 5, 6 and 7 enabled correction factors to be determined to normalise the neutron energy spectra to the position of the 1 atm counter.

All the remaining results given in this section were normalised to full power, using the ratio of activation foil measurements made by ECN staff during the second run and at full power[17]. The neutron fluence-rate spectrum of the beam in the energy range from 10 keV to 15 MeV is shown in Fig. 2. The corresponding kerma-rate spectrum is given in Fig. 3. These results are

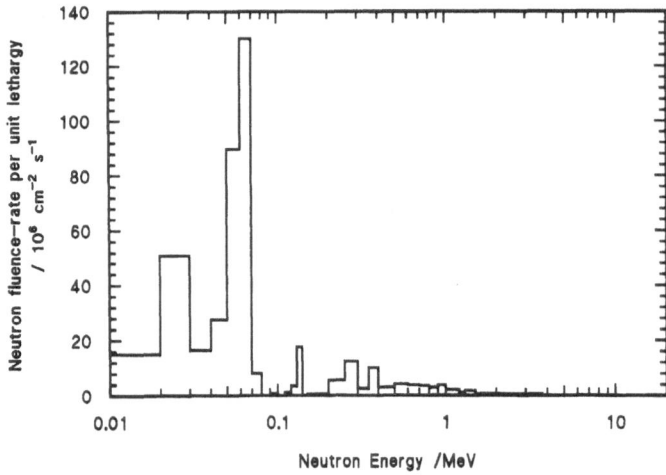

Figure 2. *Neutron fluence-rate spectrum measured with proton and alpha-recoil counters, normalised to full reactor power (45 MW, thermal), of the HFR Petten HB11 filtered neutron beam.*

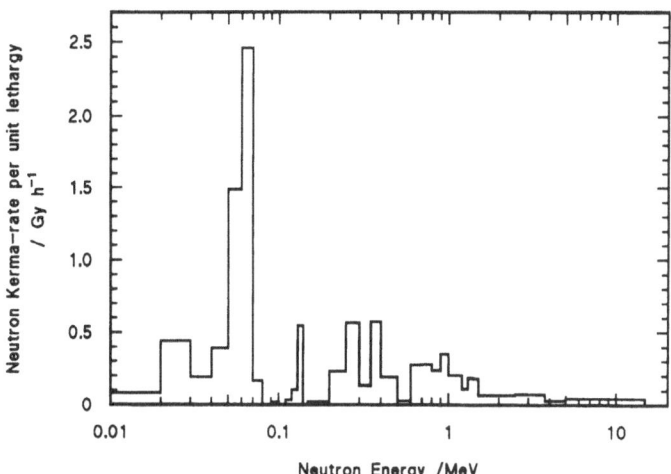

Figure 3. *Neutron kerma-rate spectrum measured with proton and alpha-recoil counters, normalised to full reactor power (45 MW, thermal), of the HFR Petten, HB11 filtered neutron beam.*

also given numerically in Table 3. In the energy range measured (10 keV to 15 MeV), the neutron fluence rate is $91.1 \cdot 10^6$ cm^{-2} s^{-1}; this corresponds to a kerma-rate in water of 1.70 Gy h^{-1}.

DISCUSSION

In the spectrum shown in Fig. 2, there are a number of peaks, particularly at 20—30 keV, 50—70 keV and 130—140 keV. The peak at 50—70 keV is attributable to the window in the argon cross-section, the others are attributable to windows in the aluminium cross-section.

Normally, electronics with simultaneous data acquisition is used to accumulate the pulse-height distributions for all the counters in the shortest time. However, it is normal to take enough spare apparatus to ensure that if any item becomes faulty, either in transportation or during the measurements, a replacement unit is available. This meant that it was possible to run the 1 atm counter separately and, hence the low level discriminator was set as low as possible for this counter enabling the low energy limit for the neutron spectrum measurement to be reduced to 10 keV. This technique proved extremely successful and should be adopted for all future measurements.

In addition to the measurements made at low power, activation detector measurements were made at full power by teams from INEL, Idaho Falls and ECN, Petten[18]. The agreement between these measurements and those de-

Table 3

Neutron energy spectrum of the HFR Petten, HB11 filtered neutron beam,
measured with proton and alpha-recoil counters

E_{min}	E_{max}	Fluence	Fluence /lethargy	Kerma-rate (in water)[16]	Kerma-rate	Kerma-rate per unit lethargy
/MeV	/MeV	/cm^{-2} s^{-1}	/cm^{-2} s^{-1}	/Gy h^{-1}	/%	/Gy h^{-1}
0.01	0.02	1.0455E+07	1.5083E+07	5.7523E—02	3.39	8.2988E—02
0.02	0.03	2.0634E+07	5.0889E+07	1.7914E—01	10.56	4.4182E—01
0.03	0.04	4.7436E+06	1.6489E+07	5.5032E—02	3.24	1.9129E—01
0.04	0.05	6.1595E+06	2.7603E+07	8.7234E—02	5.14	3.9093E—01
0.05	0.06	1.6354E+07	8.9699E+07	2.7114E—01	15.98	1.4872E+00
0.06	0.07	2.0108E+07	1.3044E+08	3.7911E—01	22.35	2.4593E+00
0.07	0.08	1.0700E+06	8.0134E+06	2.2313E—02	1.32	1.6710E—01
0.08	0.09	3.1915E+01	2.7097E+02	0.0000E+00	0.00	0.0000E+00
0.09	0.10	7.3019E+04	6.9304E+05	1.7945E—03	0.11	1.7032E—02
0.10	0.11	1.5602E+02	1.6369E+03	5.4878E—06	0.00	5.7578E—05
0.11	0.12	9.8287E+04	1.1296E+06	2.7439E—03	0.16	3.1535E—02
0.12	0.13	2.7938E+05	3.4904E+06	8.2152E—03	0.48	1.0264E—01
0.13	0.14	1.3071E+06	1.7638E+07	4.0379E—02	2.38	5.4487E—01
0.14	0.15	4.4157E+02	6.4002E+03	1.6463E—05	0.00	2.3862E—04
0.15	0.20	1.6488E+05	5.7312E+05	5.9378E—03	0.35	2.0640E—02
0.20	0.25	1.2375E+06	5.5459E+06	5.1470E—02	3.03	2.3066E—01
0.25	0.30	2.2644E+06	1.2420E+07	1.0287E—01	6.06	5.6422E—01
0.30	0.35	4.0467E+05	2.6252E+06	2.0689E—02	1.22	1.3421E—01
0.35	0.40	1.3528E+06	1.0131E+07	7.6599E—02	4.51	5.7364E—01
0.40	0.50	6.7775E+05	3.0373E+06	4.3035E—02	2.54	1.9286E—01
0.50	0.60	7.7273E+05	4.2383E+06	5.0323E—03	0.30	2.7601E—02
0.60	0.70	5.9965E+05	3.8901E+06	4.2410E—02	2.50	2.7512E—01
0.70	0.80	4.8828E+05	3.6567E+06	3.7180E—02	2.19	2.7844E—01
0.80	0.90	3.4074E+05	2.8929E+06	2.7834E—02	1.64	2.3632E—01
0.90	1.00	3.9908E+05	3.7877E+06	3.6796E—02	2.17	3.4924E—01
1.00	1.10	2.0160E+05	2.1152E+06	1.9460E—02	1.15	2.0417E—01
1.10	1.20	1.8511E+05	2.1274E+06	1.7791E—02	1.05	2.0447E—01
1.20	1.30	8.7728E+04	1.0960E+06	8.7530E—03	0.52	1.0935E—01
1.30	1.40	1.3202E+05	1.7814E+06	1.3621E—02	0.80	1.8380E—01
1.40	1.50	(1) 1.1822E+05	1.7135E+06	1.2496E—02	0.74	1.8112E—01
1.50	2.50	(1) 2.8500E+05	5.6000E+05	3.4060E—02	2.01	6.6676E—02
2.50	3.75	(1) 1.3360E+06	5.2000E+05	3.0320E—02	1.79	7.4778E—02
3.75	5.00	(1) 3.2000E+03	1.7000E+04	8.3000E—04	0.05	2.8851E—03
5.00	14.70	(1) 2.3400E+02	2.0000E+04	4.7300E—03	0.28	4.3860E—03
>14.7		(1) 0.0000E+00	0.0000E+00	0.0000E+00	0.00	0.0000E+00
Totals		9.1131E+07		1.6966E+00	100.00	

Note: (1) Values estimated from ^{4}He counter measurements

scribed in this report is very good[19]. For example, the spectrum determinedby INEL gives the neutron fluence-rate in the energy range from 10 keV to 15 MeV as $92.5 \cdot 10^6$ cm^{-2} s^{-1}, corresponding to a neutron kerma-rate in tissue of 1.476 Gy h^{-1}. This compares to a neutron fluence-rate of $91.1 \cdot 10^6$ cm^{-2} s^{-1} and neutron kerma-rate in tissue of 1.59 Gy h^{-1} determined from the proton and alpha-recoil counters. It should be noted that whilst the neutron fluence-rate in the range covered by the recoil counters is only one fifth of the total, the neutron kerma-rate in this energy range is over 92% of the total. This highlights the importance of making high resolution measurements in this energy range.

CONCLUSIONS

Neutron spectrometry measurements in the energy range from 10 keV to 15 MeV of the HFR Petten, HB11 filtered neutron beam have been described in this paper. Further details, including the description of some other subsidiary measurements, for example the beam profile at the measurement position, will be given in a future paper[8]. As with previous measurements of the HB7 beam[6], self shielding of the counters was negligible. However, small corrections were applied to take into account the variation of the neutron fluence-rate with distance from the reactor to normalise all counter measurements to the position of the 1 atm counter in Runs 1 and 2. A novel method for analysing the alpha recoil counter measurements to derive the neutron energy spectrum above 1.5 MeV was used, because of the low number of accumulated counts. A future measurement campaign might include a measurement at higher power with this counter to obtain better counting statistics. In the energy range measured (10 keV to 15 MeV), the neutron fluence rate is $91.1 \cdot 10^6$ cm^{-2} s^{-1}, this corresponds to 1.70 Gy h^{-1}. Comparison with spectra determined from activation detector measurements (made at full power) show that there is good agreement. In addition it should be noted that whilst only about one fifth of the neutron fluence-rate is in the energy range measured by the proton and alpha-recoil counters, this corresponds to over 92% of the neutron kerma-rate. This highlights the importance of measuring this part of the spectrum with high resolution.

ACKNOWLEDGEMENTS

This work was funded by the UK Department of Health. We are grateful to the HFR operators for their cooperation and for maintaining the reactor power manually during the measurement programme. We are also grateful to the support staff of ECN, Petten, particularly Mr K. Ravensburg for setting up and helping during the measurements.

REFERENCES

1. Perks, C.A.; Harrison, K.G.; Birch, R.; Delafield, H.J. The characteristics of a high intensity 24 keV iron-filtered neutron beam. *Radiat. Prot. Dosim.* 15, 31-40 (1986).

2. Perks, C.A.; Mill, A.J.; Constantine, G.; Harrison, K.G.; Gibson, J.A.B. A review of boron neutron capture therapy (BNCT) and the design and dosimetry of a high-intensity, 24 keV, neutron beam for BNCT research. *Br. J. Radiol.* 61, 1115-1126 (1988).

3. Perks, C.A.; Constantine, G.; Birch, R. The design and dosimetry of an Al/S/Ar filtered neutron beam. *Radiat. Prot. Dosim.* 23, 329-332 (1988).

4. Moss, R.L. Current overview and on the approach of clinical trials at Petten. In: *Boron Neutron Capture Therapy: Toward Clinical Trials of Glioma Treatment.* Gabel, D.; Moss, R.L., eds., Plenum Press, New York, 33 - 46 (1992).

5. Constantine, G.; Watkins, P.R.D.; Perks, C.A.; Delafield, H.J.; Ross, D.; Voorbraak, W.P.; Paardekooper, A.; Freudenreich, W.E.; Stecher-Rasmussen, F.; Moss, R.L. Progress in neutron beam development at the HFR Petten (feasibility study for a BNCT facility). In: *Proc. 8th ASTM-EURATOM Symposium on Reactor Dosimetry, Strasbourg, France, August 27-13, 1990*; accepted for publication.

6. Constantine G.; Perks, C.A.; Delafield, H.J.; Ross, D. Neutron spectrum characterisation of the HB7 beam in the HFR, Petten. *AEA Environment and Energy Report number AEA-EE-0224* (1991).

7. Watkins, P. Review of the physics calculations performed for the BNCT facility at the HFR Petten. In: *Boron Neutron Capture Therapy: Toward Clinical Trials of Glioma Treatment.* Gabel, D.; Moss, R.L., eds., Plenum Press, New York, 47 - 58 (1992).

8. Perks, C.A.; Delafield, H.J.; Constantine, G.; Stecher-Rasmussen, F.; Watkins, P.R.D. Neutron spectrometry measurements of the Petten HFR, HB11 intermediate energy neutron beam, June 1991. *AEA Environment and Energy Report*, in preparation.

9. Delafield, H.J.; Birch, R. Development and calibration of large volume proton-recoil counters for neutron spectrometry in radiological protection. *HMSO Report, AERE-R 13010* (1988).

10. Delafield, H.J.; Birch, R. Neutron spectrometry measurements with large volume cylindrical proton-recoil counters developed for use in radiological protection. *HMSO Report, AERE-R 13103* (1989).

11. Birch, R. An alpha-recoil proportional counter to measure neutron energy spectra between 2 MeV and 15 MeV. *HMSO, AERE-R 13002* (1988).

12. Birch, R.; Delafield, H.J., Perks, C.A. Measurement of the neutron spectrum inside the containment building of a PWR. *Radiat. Prot. Dosim.* 23, 281 (1988).

13. Birch, R.; Peaple, L.H.J.; Delafield, H.J. Measurement of neutron spectra with hydrogen counters. Part I. Spectrometry system and calibration. *HMSO Report, AERE-R 11397* (1984).
14. Birch, R.; Marshall, M.; Peaple, L.H.J. Measurement of neutron spectra with hydrogen counters. Part II. Analysis of proton recoil distributions. *HMSO. AERE-R 11398* (1984).
15. Siefert, A.; Casado, J.; Moss, R.L.; Gavin, P.; Philipp, K.; Huiskamp, R.; Dühmke, E. Healthy tissue tolerance studies for BNCT at the high flux reactor in Petten - First results. In: *Boron Neutron Capture Therapy: Toward Clinical Trials of Glioma Treatment.* Gabel, D.; Moss, R.L., eds., Plenum Press, New York, 179 - 188 (1992).
16. ICRU Neutron Dosimetry for Biology and Medicine. *ICRU Report 26, ICRU Publications, Washington DC* (1977).
17. Watkins, P.R.D. (1991). Private Communication.
18. Stecher-Rasmussen, F.; Constantine, G.; Freudenreich, W.; de Haas, H., Moss, R.L.; Paardekooper, A.; Ravensberg, K.; Verhagen, H.; Voorbraak, W.; Watkins, P.R.D. From filter installation to beam characterization. In: *Boron Neutron Capture Therapy: Toward Clinical Trials of Glioma Treatment.* Gabel, D.; Moss, R.L., eds., Plenum Press, New York, 59 - 77 (1992).
19. Perks, C.A.; Gibson, J.A.B. Neutron dosimetry and spectrometry for boron neutron capture therapy. In: *Proc. 7th Symposium on Neutron Dosimetry, Berlin, 14-18 October 1991* (to be published).

A SEMI-EMPIRICAL METHOD OF TREATMENT PLANNING

FOR BORON NEUTRON CAPTURE THERAPY

Cornelis P.J. Raaijmakers[a], Luc Dewit[a],
Mark W. Konijnenberg[a], Ben J. Mijnheer[a],
Raymond L. Moss[b], and Finn Stecher-Rasmussen[c]

[a]The Netherlands Cancer Institute
NL-1066 CX Amsterdam, The Netherlands

[b]Commission of the European Communities
Institute for Advanced Materials
Joint Research Centre
NL-1755 ZG Petten, The Netherlands

[c]Netherlands Energy Research Foundation (ECN)
NL-1755 ZG Petten, The Netherlands

INTRODUCTION

In BNCT the total dose delivered to the tumor and to the healthy tissue depends on several factors. Knowledge of the [10]B distribution, the thermal neutron fluence, the epithermal neutron fluence, the fast neutron fluence and the photon dose is necessary for the calculation of the total dose delivered to all points of interest in the patient. Calculations using either Monte Carlo or deterministic codes, both commonly applied in reactor physics for neutron and photon transport calculations, have been proposed for treatment planning of BNCT[1,2]. Due to the large calculation times needed for these kinds of calculations, such a procedure for external photon beam treatment planning is not yet available for daily usage in radiotherapy institutions[3]. Therefore, little experience has been gained with these procedures in clinical situations. External photon beam treatment planning is currently based on empirical knowledge of the dose distributions under reference conditions. Beam parameters measured in a large cubical water-phantom are corrected with deterministic algorithms to calculate dose distributions in other geometries[3].

Boron Neutron Capture Therapy, Edited by D. Gable
and R. Moss, Plenum Press, New York, 1992

Treatment planning of BNCT based on Monte Carlo calculations as well as using such a semi-empirical approach is in development in the Petten-Amsterdam BNCT group[4]. It is the purpose of this work to investigate wether the relatively simple semi-empirical approach can be used for the treatment planning of BNCT using an epithermal neutron beam.

EXPERIMENTAL

The epithermal HB11 beam at the High Flux Reactor (HFR) in Petten, which is designed for therapeutic application, was at the time of this study not available for experiments. Therefore an epithermal neutron beam has

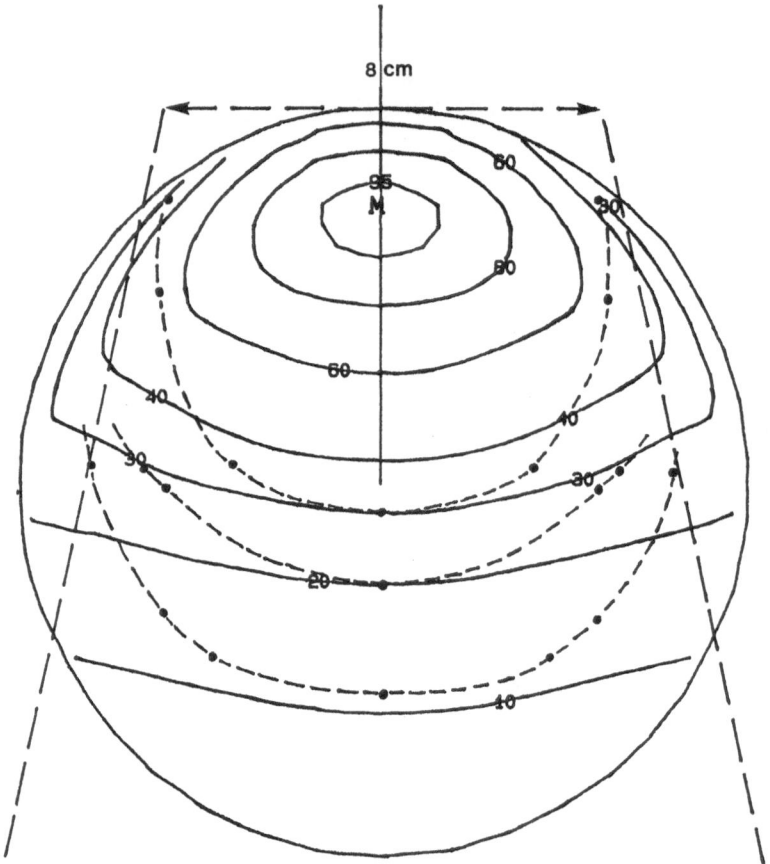

Figure 1. *Horizontal cross-section of the LFR. Beam tube HN has been adjusted to provide an epithermal neutron beam, which allows an investigation of the semi-empirical approach of treatment planning of BNCT.*

been extracted from the Low Flux Reactor (LFR), also in Petten. No effort has been put in making this beam suitable for therapy. The neutron energy spectrum is, however, a reasonable approximation of the HB11 spectrum thus allowing an investigation of the semi-empirical approach of treatment planning of BNCT. The LFR is a 30 kW nuclear reactor of the "Argonaut" type. The core of the reactor consists of 1.8 kg ^{235}U, surrounded by graphite reflectors. The neutrons are moderated by graphite. Fig. 1 shows a cross section of the LFR together with the beam tube (HN) used for these experiments. The beam has a field size of 10x10 cm^2 and a neutron fluence rate of 2.74·10^7 neutrons cm^{-2} s^{-1}. The beam is filtered by two 1 mm thick Cd-sheets, at a distance of 1.30 m. Two 10 cm long boronated polyethylene blocks with a circular opening of 8 cm diameter, placed directly behind the Cd-sheets, produce a neutron beam with a field size similar to the latest design of the HB11 beam. At the beam exit a 5 mm thick B$_4$C plate has been placed to capture thermalized neutrons. A comparison of the calculated neutron spectrum of the LFR beam and the HB11 beam shows a reasonable agreement at the low and intermediate energy side of the spectrum. The main difference between both spectra concerns the fast neutron fluence. Unlike the HB11 beam, no filtering is performed to reduce the fast neutron fluence, resulting in a larger contribution of this component to the total neutron energy spectrum. Behind the neutron collimators 5 cm thick lead blocks, with 8 cm circular openings have been placed to reduce the induced capture gamma rays. No filtering of the photon field is performed. The contribution to the dose of photons coming from the reactor core will therefore be considerably larger than the dose due to neutron capture gamma rays originating within the phantom. The inverse situation will probably occur in the HFR beam.

To investigate the spacial distribution of the different beam components in the phantoms several detectors with different shielding have been used (Table 1). A ^{235}U fission counter with an effective area of 0.8 cm^2 (Centronic FC4A/1000/-U235) is used to detect the thermal neutron fluence. The same detector covered with 1 mm Cd was used for correction of the detector response to epithermal neutrons. Photons were measured using a GM counter (Philips 18529), covered with a 3 mm ^{6}LiF cap to prevent thermal neutrons from reaching the detector, and a neutron insensitive 1.0 cm^3 magnesium ionization chamber filled with argon. Dead time corrections of the GM counter reading were applied. Calibration of the neutron fluence was done with AuAl and MnNi foils. Photon detectors were calibrated using the reference gamma-ray detector calibration set-up at Petten. Two phantoms were irradiated. A cubical phantom made of 8 mm thick Perspex plates with overall dimensions of 40x50x60 cm^3 and a cylindrical phantom with a diameter of 15 cm and a height of 30 cm made of 1.5 mm thick Perspex. Both phantoms were filled with normal water. The planning system used for these first tests was the Nucletron 2-D external photon beam treatment planning system.

Table 1

*The different detectors used for the measurements
of the different beam components*

Beam component	Detector
Thermal neutron fluence	^{235}U fission counter AuAl + MnNi foils
Epithermal neutron fluence	^{235}U fission counter + Cd AuAl + MnNi foils
Gamma-ray dose	GM counter + ^{6}LiF Mg(Ar) ionization chamber TLD
Neutron plus gamma-ray dose	TE(TE) ionization chamber

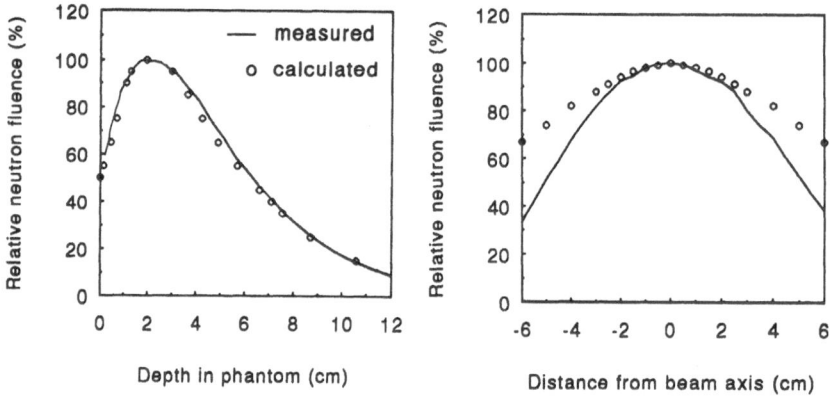

Figure 2. *a) Measured and calculated thermal neutron fluence along the beam
axis in the cylindrical phantom. b) Measured and calculated beam profiles in
the cylindrical phantom at 7 cm depth.*

RESULTS

To obtain beam characteristics under reference conditions, measurements have been carried out in the large cubical phantom using the ^{235}U fission counter. The thermal neutron fluence as a function of depth along the central beam axis as well as cross plane beam profiles at different depths have been determined. A correction for the detector response to epithermal neutrons has been performed. The beam broadening was seen to be linear inside the phantom and could be extrapolated to a virtual source-phantom distance of 20.5 cm. The beam profiles inside the phantom and the incident beam profile show little similarity. This is caused by the high scattering cross-section of hydrogen which limits the influence of beam width and field flatness of the incident beam on the thermal neutron distribution inside the phantom.

The maximum thermal neutron fluence was at 2.8 cm depth, a value typical for an epithermal neutron beam[5]. At 10 cm depth inside the phantom only 20% of this value was obtained. Due to the thickness of the Perspex phantom wall and the diameter of the detector, the first point of measurement was situated at 9.5 mm depth. By using AuAl and MnNi foils it was possible to measure the thermal neutron fluence at the surface of the phantom and to perform an absolute calibration. The peak thermal neutron fluence was determined as $1.17 \cdot 10^7$ neutrons cm^{-2} s^{-1}, while at the surface 20% of this value was measured.

The photon fluence distribution was measured using the GM counter. Due to its low sensitivity and consequently large measuring times, the Mg(Ar) ionization chamber was only used for measurements along the beam axis. In spite of the low neutron cross sections of the materials used in this chamber, its sensitivity for neutrons was somewhat larger than that of the GM counter covered with ^{6}LiF. At larger depths, where the thermal neutron fluence became small, there was good agreement between both sets of values. Subsequently the thermal neutron fluence distribution in the cylindrical phantom was investigated. Again, beam profiles and the depth distribution were determined. On the central beam axis the difference in neutron fluence distribution between both phantoms was small; only at depths larger than 10 cm could a small difference be detected. Beam profiles are seen to be less broad in the cylindrical phantom. This can be explained by the lack of scatter near the edges of the phantom, where less material is present from which thermal neutrons can be scattered, resulting in a lower thermal neutron fluence. The results obtained in the cubical reference phantom were used as input for the treatment planning system. First interpolations and iso-thermal neutron fluence curve calculations were performed. Subsequently the planning system was used to calculate from these data, the thermal neutron fluence distribution in the cylinder. Corrections were made by the program for differences in the phantom shape at the beam entrance side. No corrections were made in this treatment planning system for lack of scatter. Beam profiles at different depths, the fluence distribution along the central axis and iso-fluence curves were calculated in the central plane of the beam. Results of the calculations are shown in Fig. 3. The calculated fluence distribution along the beam axis was in good agreement with the measurements indicating that the lack of scatter is almost negligible in the middle of the phantom. A maximum difference of 5% occurred at 5 cm depth. The calculated beam profile at 2 cm depth was also in good agreement with the measured profile. No significant differences were found here, indicating that at this depth the difference between both phantoms in thermal neutron fluence is caused by differences in phantom shape at the beam entrance side. At larger depths differences between measured and calculated profiles were apparent. As can be seen in Fig. 3, the error at the edges of the phantom, caused by not taking into account the lack of scatter, is considerable. As a consequence large corrections for scatter effects are required for accurate calculation of thermal neutron fluence throughout the total volume. It should be noted that in the procedure we followed the reference phantom had a volume much larger than the phantom in which the thermal neutron fluence was calculated.

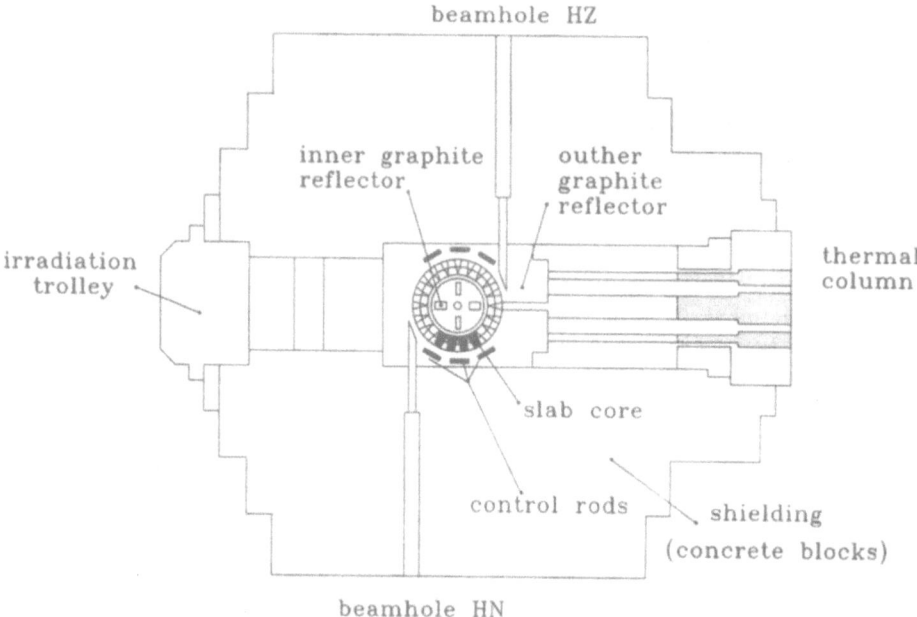

Figure 3. *Iso-fluence lines for thermal neutrons in the central plane of a cylindrical phantom as calculated by the treatment planning system (solid lines). The dashed curves show the measured iso-fluence lines at 10, 20 and 30% of the maximum value.*

CONCLUSIONS

The main difference between treatment planning of an external photon beam and an epithermal neutron beam is the large scatter component of the neutron beam. This component limits the influence of geometric characteristics of the incident beam and enlarges the influence of phantom geometry on the thermal neutron distribution inside the phantom. Adjustment of 2-D photon beam dose calculation procedures with respect to the scattered component of the beam is therefore required for a semi-empirical method of treatment planning of an epithermal neutron beam when using a large cubical phantom as reference phantom. Due to the large influence of phantom geometry, using another reference phantom that is more of a resemblance to the clinical situation, might be necessary. For brain irradiations, the first goal of the BNCT project in Petten, the cylindrical phantom seems to be a good alternative reference phantom. For dose specification purposes the simple approach presented in this paper seems to be promising.

ACKNOWLEDGEMENTS

The authors would like to thank the crew of the LFR, A. Paardekoper and W. Voorbraak for analyzing the foil measurements, H. Verhagen for his support during photon dose measurements and J.H. Lanson for helping with the treatment planning system calculations. This work was financially supported by the Dutch Cancer Society under project NKB grant NKI 90-03.

REFERENCES

1. Nigg, D.W.; Randolph, P.D.; Wheeler, F.J. Demonstration of three-dimensional deterministic radiation transport theory dose distribution analysis for boron neutron capture therapy. *Med.Phys.* 18, 43-53 (1991).

2. Deutsch, O.L.; Murray, B.W. Monte Carlo dosimetry calculation for boron neutron capture therapy in the treatment of brain tumors. *Nuclear Tech.* 25, 320-339 (1975).

3. ICRU Report 42. Use of computers in external beam radiotherapy procedures with high-energy photons and electrons. International Commission on Radiation Units and Measurements, Bethesda MD, USA (1987).

4. Konijnenberg, M.W.; Raaijmakers, C.P.J.; Dewit, L.; Mijnheer, B.J.; Moss, R.L.; Stecher-Rasmussen, F.; Watkins, P.R.D. A Monte Carlo technique for treatment planning of boron neutron capture therapy. *Radiat. Prot. Dosim.*, in the press.

5.   Fairchild, R.G.; Bond, V.P. Current status of [10]B neutron capture
      therapy: enhancement of tumor dose via beam filtration and dose
      rate, and the effects of these parameters on minimum boron content:
      a theoretical evaluation. *Int. J. Radiat. Oncol. Biol. Phys.* 11, 831-840
      (1985).

PRESENT STATUS OF THE THREE-DIMENSIONAL TREATMENT

PLANNING METHODOLOGIES FOR THE

PETTEN BNCT FACILITY

Peter Watkins

Institute for Advanced Materials
JRC Petten
NL-1755 ZG Petten, The Netherlands

INTRODUCTION

Traditional treatment planning systems for photon and fast neutron ir-radiations invariably rely on one or two dimensional computer models often with a significant number of empirical adjustments to obtain realistic results. Several aspects of BNCT mean that the use of approximate one- and two-dimensional treatment planning tools may be inadequate. In particular the epithermal neutrons entering the target region travel only very short distances before being scattered or absorbed in the medium which is very different from the situation with photons or fast neutrons. Thus an initially parallel beam diverges very rapidly within the target. The result is a "cloud" of neutrons of various energies distributed throughout the target volume.

BNCT requires a number of other quantities to be quantified by the treatment planning tool in addition to the distribution of the incident beam within the target. Dose distributions due to fast neutrons, induced gamma photons, protons from the nitrogen (n,p) reaction and the crucial $^{10}B(n,\alpha)^7Li$ reaction all have to be evaluated separately. The last of these, the $^{10}B(n,\alpha)^7Li$ dose is strongly correlated with the distribution of the thermal neutron flux. This is not a component of the incident beam but is generated within the target by transport and moderation processes of the incident epithermal neutrons. Thus to correctly represent the ^{10}B dose it is necessary to model these processes in some detail.

For a large aperture beam the position of the peak thermal flux is dependent upon the geometry of the target and it is not at all certain that a reconstruction of the full 3D flux distribution can be synthesised from 1D and 2D models.

With all of the above-mentioned uncertainties there is clearly a need to develop a treatment planning method which is capable of modelling the real situation as accurately as possible. This implies a detailed 3D model with an exact treatment of neutron and photon transport. For regular, day-to-day, treatment planning calculations a method of this sophistication may not be essential and the feasibility of using other approximate methods is currently being examined[1].

MCNP AS A TREATMENT PLANNING TOOL

In the absence of a specialised 3D treatment planning tool we have used the Monte Carlo code MCNP[2] for the particle transport calculations. The dose rates thus produced are subsequently manipulated by a separate code, TREAT, to generate total doses, irradiation times and various other data. Although not ideal MCNP does address the following points:

- It has a full 3-dimensional capability allowing complex shapes to be modelled exactly, although creating the geometry in such a general purpose code can involve a considerable effort.

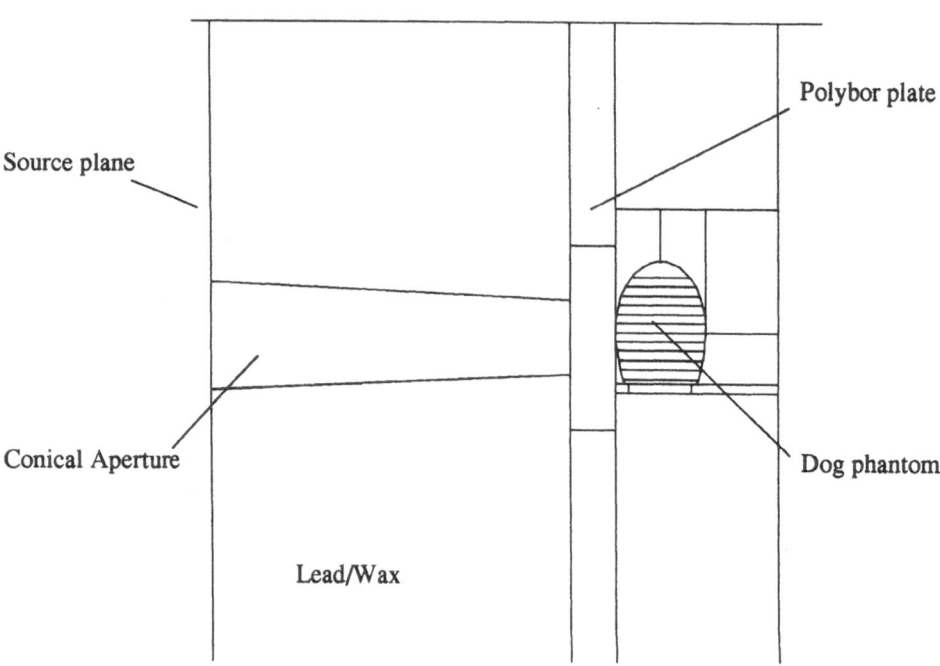

Figure 1. *MCNP Model of the Beagle Head Phantom (Side View).*

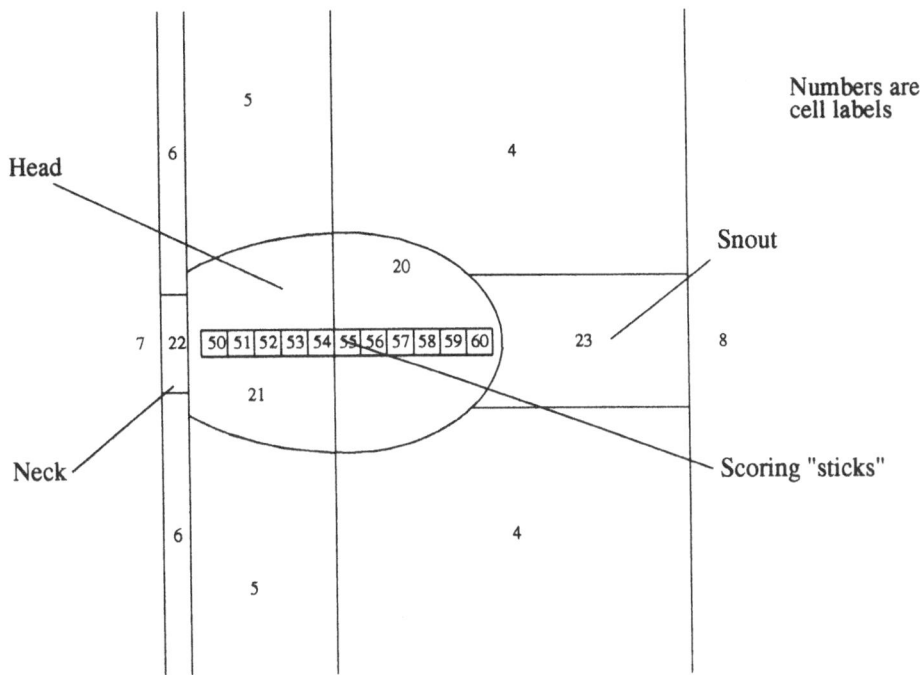

Figure 2. *MCNP Model of the Beagle Head Phantom (Plan View).*

- The detailed neutron and photon transport through the target medium is represented explicitly including the production of induced photons resulting from neutron interactions within the target.
- Continuous energy cross-section data is available in an extensive and reasonably well validated library.
- Rotation of the target region can also be modelled so multi-portal planning is possible in principle.
- Fluxes, doses, reaction rates, etc. may be tallied over a wide variety of spatial and energy regions.

BASIC MCNP MODEL

Calculated values of the neutron spectrum and intensity from the HB11 design phase[3] are used to generate a neutron source at the reactor side of the final conical aperture, see Fig. 1. Specifying the source on this particular surface permits many different apertures to be represented for a given source. However it is sufficiently removed from the target region so that any perturbations to the source caused by the presence of a target are minimal.

In addition to the neutron source there are photons within the beam which have been generated in the reactor core and in the filter assembly of the BNCT facility. These also have to be accounted for in the treatment planning system. At present these photon sources have to be calculated in a subsidiary calculation. One of the restrictions of the MCNP code is that it is incapable of performing calculations with a combined photon and neutron source.

The target region is divided (by hand) into 1cm square "sticks" extending over the full depth of the target, see Fig. 2. Each stick is divided along its length into 0.5cm sections. For each of the 1cm x 1cm x 0.5cm voxels thus specified in the target the following quantities are tallied in three energy groups:

- Neutron Flux
- Photon dose rate
- Fast neutron dose rate
- Nitrogen (n,p) dose rate
- Boron ^{10}B(n,α)^{7}Li dose rate (assuming 50 ppm boron)

For the purposes of these calculations we have assumed the following energy group structure:

Thermal	0.0	to	0.414 eV
Epithermal	0.414 eV	to	10 keV
Fast	10 keV	to	20 MeV

PROCESSING THE MCNP RESULTS

The results from the MCNP simulations form a database of dose rates and neutron fluxes for each scoring region of the target volume. The database is then interrogated interactively with a code called TREAT which extracts various requested data and evaluates **physical** doses for selected sticks or voxels. At present there is no facility to convolute these data with RBE values although this is a trivial extension and will be incorporated in the near future.

TREAT requires as input the actual boron concentration in ppm used in the irradiation and also the target peak (physical) dose. TREAT then scans the database of dose rates to evaluate the position of the peak total dose rate and then calculates the irradiation time that is needed to reach the target total dose. It should be noted that the position of the peak dose will vary markedly with boron concentration and even more so if RBE values are included. Thus for each set of input data, ^{10}B concentration and RBE values, it is essential that the whole database be examined to re-establish the position of the peak total dose.

For any specified stick through the target, TREAT will evaluate the various doses appropriate for the calculated irradiation time.

USE OF THE MCNP/TREAT SYSTEM

Some irradiations have already been performed at the HFR Petten using
beagle dogs and also a beagle phantom made from lucite. For all of these irradiations the MCNP/TREAT system has been used to predict the dose rates
and to establish the irradiation times. As an illustration the beagle phantom
experiment is described here in more detail. The MCNP model for this is
shown in Figs. 1 and 2. The head phantom is modelled in two parts, two quadratic surfaces joined at a central plane. In addition the snout and neck are
represented approximately. The phantom is solid, with no voided regions
within it to model the mouth or other cavities. Some 11 "sticks" were
specified, Fig. 2, through the phantom to score the fluxes and doses. These
were divided along their lengths into 0.25 and 0.5 cm cuboids. In all some 150
scoring regions were defined but, for this calculation, only along the sagittal
plane.

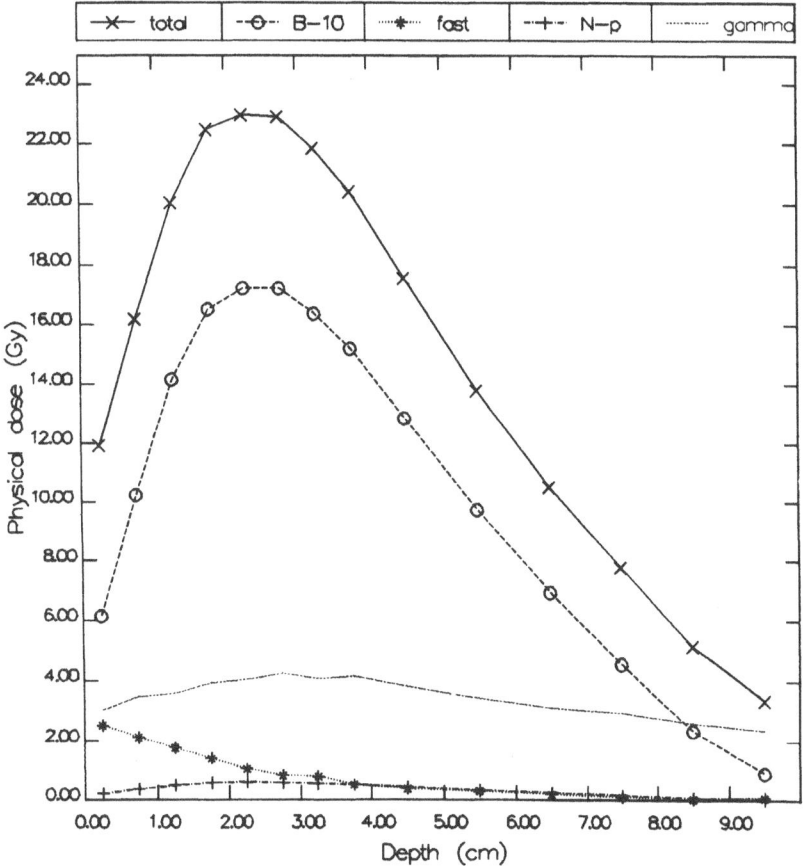

Figure 3. *Dose components along a stick as predicted by MCNP/TREAT.*

Table 1

Treatment Planning Results from MCNP/TREAT
Sample output

Extracting results for cell 54 from file dog05out
<u>Summary of results for cell 54</u>

Maximum thermal flux is 6.69489e+08 in cell 54 at x= 2.75

Depth	Thermal Flux	< ---------------- Dose Rates (Gy/hr) ---------------- >					
(cm)	(n cm^{-2} s^{-1})	B-10	N(n,p)	Fast	Gamma	Incident	Total
0.250	2.34508e+08	3.148	0.110	1.293	0.554	1.000	6.104
0.750	4.01515e+08	5.233	0.180	1.088	0.779	1.000	8.281
1.250	5.47165e+08	7.242	0.249	0.911	0.849	1.000	10.251
1.750	6.45278e+08	8.447	0.291	0.730	1.014	1.000	11.482
2.250	6.66287e+08	8.807	0.304	0.545	1.071	1.000	11.726
2.750	6.69489e+08	8.807	0.303	0.424	1.172	1.000	11.705
3.250	6.24928e+08	8.376	0.289	0.415	1.094	1.000	11.175
3.750	5.77706e+08	7.771	0.268	0.265	1.129	1.000	10.433
4.500	4.92694e+08	6.591	0.227	0.195	0.962	1.000	8.975
5.500	3.72546e+08	4.975	0.171	0.147	0.758	1.000	7.051
6.500	2.65057e+08	3.555	0.123	0.100	0.604	1.000	5.382
7.500	1.73278e+08	2.335	0.080	0.047	0.521	1.000	3.983
8.500	9.07542e+07	1.219	0.042	0.021	0.348	1.000	2.630
9.500	3.63688e+07	0.470	0.016	0.032	0.215	1.000	1.733

Peak dose-rate with 50 ppm boron is 11.75 Gy/hour in cell 53 at x= 2.25
Peak surface dose-rate with 50 ppm boron is 6.10 Gy/hour in cell 54
Time for peak dose of 23.00 Gy with 50 ppm boron is 1.96 hours

<u>Doses in cell 54 after 1.96 hours with 50 ppm boron</u>

Depth	< ---------------- Doses (Gy) ---------------- >					
(cm)	B-10	N(n,p)	Fast	Gamma	Incident	Total
0.250	6.160	0.214	2.529	1.084	1.957	11.945
0.750	10.240	0.353	2.130	1.525	1.957	16.205
1.250	14.172	0.488	1.782	1.661	1.957	20.060
1.750	16.529	0.570	1.429	1.985	1.957	22.469
2.250	17.233	0.595	1.066	2.095	1.957	22.946
2.750	17.233	0.593	0.830	2.293	1.957	22.906
3.250	16.391	0.565	0.813	2.141	1.957	21.867
3.750	15.206	0.525	0.518	2.209	1.957	20.415
4.500	12.898	0.444	0.382	1.882	1.957	17.563
5.500	9.735	0.336	0.287	1.483	1.957	13.798
6.500	6.957	0.240	0.195	1.183	1.957	10.531
7.500	4.569	0.157	0.092	1.020	1.957	7.795
8.500	2.385	0.082	0.041	0.681	1.957	5.147
9.500	0.919	0.032	0.062	0.422	1.957	3.391

The output from the TREAT code is shown in Table 1 for one selected cell (stick). Here the first part of the output shows the various dose rates together with the thermal flux. There are two photon dose rates indicated; the column labelled "Gamma" is that induced by neutron interactions within the phantom, whilst "Incident" represents the gamma component incident on the phantom. At present this is assumed to be a constant and is an input parameter for TREAT. In the future the incident photon dose will be evaluated in a separate MCNP simulation and then included as a second set of data for TREAT. The required peak physical dose is also input and TREAT evaluates the irradiation time and the different dose components. These are shown graphically in Fig. 3.

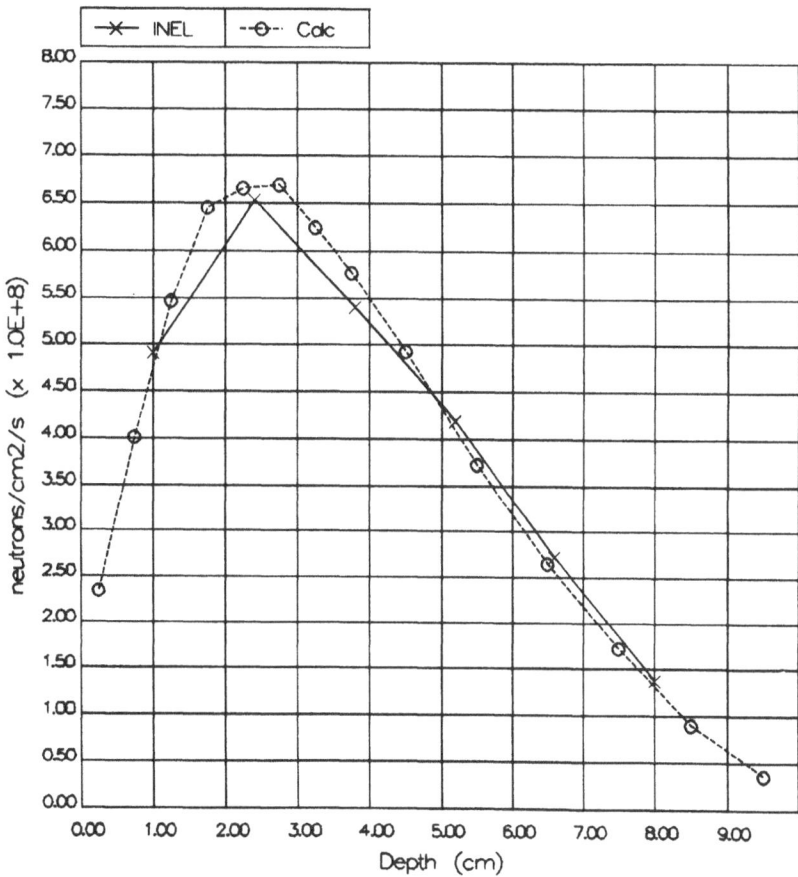

Figure 4. *Comparison of measured and calculated thermal fluxes in the beagle phantom.*

This particular case of the lucite dog phantom also allowed us to make a comparison with measured data. A number of fine holes have been drilled into the lucite into which fine dosimetry wires and TLDs may be inserted. Using this approach the INEL foil irradiations[4] have provided estimates of the thermal flux at various depths within the phantom. The measured values are compared in Fig. 4 with those estimated by the MCNP/TREAT system. The agreement is seen to be very good.

DEFICIENCIES OF THE PRESENT MODEL

There are a number of inadequacies with this system which make it rather cumbersome to use in its present state. In general these are problems associated with using a general purpose Monte Carlo code such as MCNP for what is really a very specialised task.

Creating the geometrical model within the MCNP simulation is a time consuming job and is restricted to the specification of surfaces that MCNP understands. These can be fairly complex but may be inadequate for specifying the detailed shape of a dog's head for example. The target volume itself then has to be divided into voxels (or sticks) for scoring the dose rates and fluxes. Again this must be performed at some length by hand. It is also necessary to have available all the volumes of the voxels - a non-trivial exercise if they happen to intersect one of the complex boundary curves of the target.

Having created the MCNP model the actual Monte Carlo simulation is run. This may take a number of hours to complete using present day workstations and computers. Again there would be considerable improvements if a specialised code were to be used. However computers are rapidly becoming faster and it would already be possible to reduce run times to minutes if, for example, a transputer array were to be used for this task.

One serious drawback to the use of MCNP is its inability, at present, to perform calculations from a combined neutron and photon source. As a result it is necessary to perform 2 separate calculations from a neutron and photon source respectively and then combine the data at a later stage. It is to be hoped that this problem will be overcome in future versions of MCNP.

The combination of various sets of dose rate data is a task that is performed by the TREAT code. Multi-portal irradiations could be modelled by performing an MCNP simulation for each portal in turn and generating dose rate distributions on a common scoring mesh within the target volume. TREAT would then combine these different sets of dose rates as required to produce a final dose distribution. This concept requires further development.

Although the MCNP/TREAT system works effectively as indicated by the example presented in this paper it is by no means a "user friendly" system at the present time. In particular several interfaces to graphics displays have yet to be constructed which will permit a smoother and simpler operation.

CONCLUSIONS

The reasons for developing a sophisticated 3D treatment planning tool for BNCT have been explained. A possible solution using the Monte Carlo code MCNP coupled with a data handling code TREAT is under development. An example of the use of this system for irradiations within the Petten healthy tissue tolerance study has been described. The results that have been obtained agree well with measured data.

The system requires a number of extensions and improvements before it can be used on a regular basis for BNCT planning. However the initial work has been very promising.

REFERENCES

1. Raaijmakers, C.P.J.; Dewit, L.; Konijnenberg, M.W.; Mijnheer, B.J.; Moss, R.L.; Stecher-Rasmussen, F. A Semi-empirical method of treatment planning for boron neutron capture therapy. In: *Boron Neutron Capture Therapy: Toward Clinical Trials of Glioma Treatment.* Gabel, D.; Moss, R.L., eds., Plenum Press, New York, 93 - 100 (1992).
2. Briesmeister, J.F.(Ed.), MCNP - a General Monte Carlo Code for Neutron and Photon Transport, September 1986.
3. Watkins, P.; Konijnenberg, M.; Constantine, G.; Rief, H.; Ricchena, R.; de Haas, J.B.M.; Freudenreich, W. Review of the physics calculations performed for the BNCT facility at the HFR Petten. In: *Boron Neutron Capture Therapy: Toward Clinical Trials of Glioma Treatment.* Gabel, D.; Moss, R.L., eds., Plenum Press, New York, 47 - 58 (1992).
4. Harker, Y.D., Amaro, C.R; Watkins, P.R.D.; Voorbraak, W.P. Neutron and gamma measurements at the Petten High Flux Reactor (HB-11) beam. In: *Proc. of 1992 ANS Annual Meeting on Neutron Capture Therapy.* Boston, MA (1992).

A PHASE 1 BIODISTRIBUTION STUDY OF

p-BORONOPHENYLALANINE

Jeffrey A. Coderre

Medical Department
Brookhaven National Laboratory
Upton, NY 11973, USA

INTRODUCTION

The United States Food and Drug Administration (FDA) has approved an Investigational New Drug (IND) Exemption for the boron-containing amino acid p-boronophenylalanine (BPA). The IND application was based upon pre-clinical studies carried out at Brookhaven National Laboratory that included: 1) the demonstration of selective accumulation of boron in a murine melanoma[1]; 2) the report of successful boron neutron capture therapy in a murine melanoma following BPA administration[2]; and 3) the toxicology/histopathology reports following oral administration of BPA to mice and rabbits[3]. The IND application also included a review and bibliography of the Japanese results with BPA both in animals and in humans. Originally approved for studies in patients with melanoma, the BPA IND was amended to include patients with glioblastoma or breast cancer following the report that BPA selectively delivered boron to tumors other than melanoma[3].

The IND Sponsor is A. Meek, MD, Chairman, Department of Radiation Oncology, State University of New York, Stony Brook, NY. Other participating investigators in the BPA Phase 1 biodistribution study include:

1. T. Nowak, MD, Department of Neurosurgery, State University of New York, Stony Brook, NY.
2. S. Packer, MD, Chief of Ophthalmology, Department of Ophthalmology, North Shore University Hospital, Manhasset, NY.
3. D. Wazer, MD, R. Zamenhof, PhD, Department of Therapeutic Radiology, and S. Saris, MD, Department of Neurosurgery, New England Medical Center, Boston, MA.

Boron Neutron Capture Therapy, Edited by D. Gable
and R. Moss, Plenum Press, New York, 1992

4. G. Rogers, MD, Department of Dermatology, University Hospital, Boston, MA.
5. R. Gahbauer, MD, Chairman, Department of Radiation Oncology, Ohio State University, Columbus, OH.
6. Z. Fuks, MD, A. Houghton, MD, Department of Radiation Oncology, Memorial Sloane-Kettering Cancer Center, New York, NY.
7. J. Coderre, PhD, Medical Department, Brookhaven National Laboratory, Upton, NY.

The objectives of the Phase 1 BPA biodistribution study are as follows:

Objective 1

To establish the safety of orally administered BPA as determined by monitoring of patients' vital signs and by clinical analysis of blood before and after BPA administration.

Objective 2

To establish BPA pharmacokinetics by monitoring the rates of boron absorption into and clearance from the blood and the rate of urinary excretion of boron.

Objective 3

To measure the amount of boron incorporated into human tumors (melanoma, glioma, and breast carcinoma) using samples obtained at surgery or biopsy.

This report presents the results obtained from the first thirteen patients entered into the study. Three additional glioblastoma patients have been studied recently at Stony Brook, the tissues are still being analyzed.

MATERIALS AND METHODS

BPA

The [10]B-enriched L-BPA used for the first four patient studies was prepared at the Medical Department, Brookhaven National Laboratory. The BPA synthesis was carried out as described by Snyder[4], with some minor modifications[1]. The D and L enantiomers of BPA were resolved enzymatically as described by Roberts[5]. The Callery Chemical Company, Callery, PA, has obtained a Drug Master File with the FDA for the preparation of patient grade [10]B-enriched L-BPA. All patient studies now utilize the Callery BPA. BPA is administered to patients orally as a slurry of the crystalline, free amino acid in water or fruit juice.

Boron Analysis

Boron analysis was performed by measuring the 478 keV photons produced during $^{10}B(n,\alpha)^{7}Li$ reactions[6]. Samples up to 1.0 g can be analyzed in 200 s with an error (± 1 SD) of $\approx 15\%$ at the lower limit of detection. (≈ 1 μg ^{10}B), and errors of $\approx 10\%$ at ≈ 5 μg ^{10}B, and $\approx 5\%$ when the amount of ^{10}B in the sample is 10 μg or greater. Calibration of the prompt-gamma analytical system for ^{10}B determination was performed on each day of measurement using US National Bureau of Standards ^{10}B-enriched boric acid. Patient samples from studies carried out at the New England Medical Center or at the Boston University Hospital were analyzed at the Massachusetts Institute of Technology by either a prompt gamma technique patterned after (and cross-calibrated with) the facility at BNL or by a high resolution track etch method[7].

RESULTS AND DISCUSSION

BPA Dose Escalation

Sixteen patients have been entered into the BPA Phase 1 biodistribution study as of September 1991. Data for the first thirteen patients have been analyzed and are included in this report Table 1 lists the tumor types studied to date. The ocular melanoma patients were from North Shore University Hospital (Dr. Packer), the glioblastoma patients were from the New England Medical Center (Dr. Saris) and the cutaneous melanoma patients were from University Hospital (Dr. Rogers). In the thirteen patients studied to date there were no BPA-related effects on patient's vital signs (monitored for two hours post-administration) or on blood clinical chemistries at any dose level. The dose escalation schedule was recommended by the staff at the FDA. To put the low initial doses in perspective, the doses routinely given to mice, rats and rabbits in studies at BNL range from 500 to 750 mg BPA/kg body weight. The doses used in the toxicity/histopathology studies submitted to the FDA utilized oral doses of BPA that ranged up to 5000 mg/kg body weight with no observable compound-related effect[3].

Blood Pharmacokinetics

For pharmacokinetic modelling, a one compartment oral absorption model was used, assuming first order absorption into and first order elimination from a volume of distribution. The volume (V) of the compartment was calculated as 65% of body weight (an estimate of total body water). The following equation was used to fit to the experimental data,

$$y = (\frac{FD}{V}) (\frac{a}{a-b}) (e^{-bt} - e^{-at})$$

where: y = the boron concentration (μg ^{10}B/g) in blood at time t
    D = The administered dose in mg ^{10}B (given orally in a single dose)
    V = the volume of distribution (total body water, 65% of body weight)
    F = the absorbed fraction of the administered dose
    a = the first order rate constant for absorption into V
    b = the first order rate constant for elimination from V.

Table 1

Patient Information

Patient	Tumor Type	Dose mg BPA/kg	Dose mg BPA/m^2
VZ1	ocular melanoma	27	1000
BW2	ocular melanoma	54	2000
LR3	ocular melanoma	54	2000
MS4	ocular melanoma	54	2000
ES5	ocular melanoma	90	3340
LF6	cutaneous melanoma	90	3340
JR7	glioblastoma	90	3340
JM8	glioblastoma	135	5010
ES9	cutaneous melanoma	135	5010
RB10	cutaneous melanoma	135	5010
SZ11	ocular melanoma	189	7014
JB12	ocular melanoma	189	7014
SN13	cutaneous melanoma	189	7014

The blood boron concentrations, y, were determined by the prompt gamma technique; D, and V were based upon the individual patient's weight. The variables F, a, and b were fit to the experimental data for each patient by successive iteration using the curve fitting function in the graphics software package SigmaPlot 4.0. Table 2 lists the patient weight, the parameters V and D as well as the variables F, a, and b derived from the curve fitting for all thirteen patients. Also shown in Table 2 are the half times for absorption of ^{10}B into ($t_{1/2}$ abs) and elimination of ^{10}B from ($t_{1/2}$ elim) the volume of distribution, V. The last row in Table 2 provides the mean $\pm$ SD for the pharmacokinetic parameters and variables. As an example, Figure 1 shows the blood data and the plot of the fitted curve for patient SN13.

An important variable that will be monitored closely as the BPA biodistribution study proceeds is F, the fraction of the oral dose actually absorbed from the gastrointestinal tract. In animal tumor models (mice, rats and rabbits) as the amount of BPA contained in a single oral dose was increased, a point was reached beyond which no additional accumulation of ^{10}B in the tumor was observed (J. Coderre, unpublished data). It was inferred that the absorptive ability of the gastrointestinal tract was saturated. However, two properly timed oral doses have proven to be additive with respect to loading of the tumor with ^{10}B, while maintaining the same tumor-to-normal-tissue ^{10}B ratios[8]. With the exception of patient SZ11, the F values in Table 2 appear to be decreasing as the BPA dose increases. If the trend continues, alternative dosing schedules will be evaluated.

Table 2

Blood Pharmacokinetic Parameters

Patient	wt. (kg)	V (l)	D (mg ^{10}B)	F (%)	a (hr^{-1})	b (hr^{-1})	$t_{1/2}$ abs (hr)	$t_{1/2}$ elim (hr)
VZ1	81.0	52.7	107.2	51.18	1.00	0.05	0.69	14.87
BW2	95.0	61.8	251.4	74.12	0.18	0.18	3.87	3.84
LR3	86.0	55.9	227.6	60.07	1.60	0.04	0.43	16.31
MS4	81.8	53.2	216.4	47.19	0.75	0.08	0.92	8.37
ES5	45.0	29.3	198.4	58.55	0.58	0.09	1.19	7.41
LF6	64.5	41.9	284.4	35.80	0.81	0.05	0.85	14.56
JR7	91.0	59.2	401.3	36.60	0.40	0.08	1.72	8.49
JM8	75.0	48.8	496.1	16.71	0.44	0.04	1.56	16.46
ES9	63.6	41.3	420.7	26.18	0.92	0.04	0.75	15.61
RB10	70.4	45.8	465.7	23.28	0.81	0.05	0.86	15.10
SZ11	70.9	46.1	656.6	58.51	0.22	0.09	3.17	7.57
JB12	62.3	40.5	577.0	38.28	0.40	0.07	1.74	9.28
SN13	93.0	60.5	861.3	20.41	0.53	0.03	1.31	23.26
mean	75.3	49.0		42.07	0.66	0.07	1.47	12.39
±SD	±14.5	±9.4		±17.8	±0.38	±0.04	±1.01	±5.33

Urine Pharmacokinetics

Figure 2 shows the cumulative excretion of ^{10}B in urine as a function of time for the first five patients studied. Each sample was collected separately and the boron content analyzed at BNL by the prompt gamma method. The fraction of the administered dose (in mg ^{10}B) recovered in the urine within 48 hours was 61% for patient VZ1 (open squares), 34% for patient BW2 (closed triangles), 42% for patient LR3 (open triangles), 19% (only 19 hours collection) for patient MS4 (closed circles), and 65% for patient ES5 (open circles). Urine was collected in two 24-hour pools (not graphed) from patients SZ11 and JB12; the fractions of the administered dose recovered were 38% and 41%, respectively. These values for the amount of ^{10}B recovered in urine are in relatively good agreement with the values for F, the fraction of the administered dose absorbed, derived from the blood data (Table 2).

Uptake of ^{10}B in Human Tumors

Table 3 lists the time between BPA administration and surgery, the concentration of ^{10}B in tumor and the concentration of ^{10}B in blood at the time of surgery for the thirteen patients studied to date. Figure 3 shows the ^{10}B concentrations in tumor samples from patients that received 90 mg BPA/kg or more (patients 6-13 in the study, see Table 3). For three patients, multiple samples were analyzed from a single surgical specimen; all data points are plotted. Thus, the three inverted triangles at 6.5 hrs are from patient RB10, the two closed circles at 8 hrs are from patient JR7 and the two closed circles at 9.5 hrs are from patient JM8. The blood simulation line was obtained by

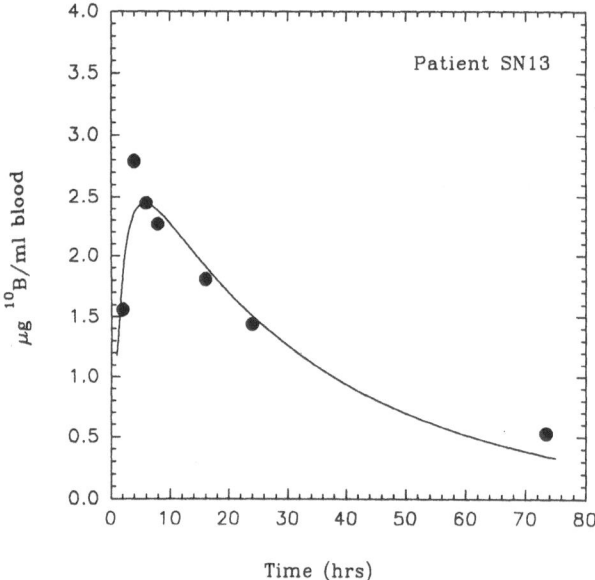

Figure 1. ^{10}B concentration (μg $^{10}B/g$) in blood as a function of time following BPA administration (data points). The solid line is the fit of a one compartment oral absorption pharmacokinetic model to the experimental data (see text). The data shown are an example for one patient (SN13) to illustrate the rapid uptake of boron into ($t_{1/2}$ = 1.3 hours) and slower clearance of boron from ($t_{1/2}$ = 23.3 hours) the blood.

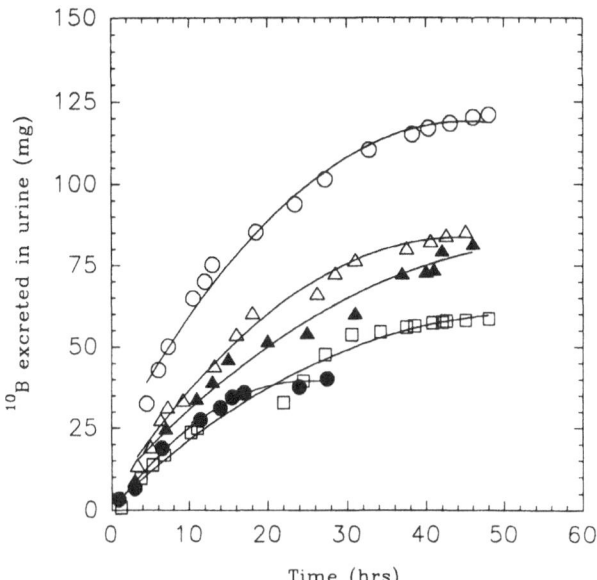

Figure 2. *Cumulative excretion of* ^{10}B *in urine versus time post-administration for patients VZ1 (open squares), BW2 (filled triangles), LR3 (open triangles), MS4 (filled circles), and ES5 (open circles). The data are plotted as the running total (each new point contains the sum of all previous points).*

plotting the solution to the one compartment oral absorption model using the mean values for F, V, a and b obtained from the fitting of all patient's individual blood data (see Table 2), and 189 mg/kg for D. The actual tumor-to-blood ^{10}B concentration ratios for each individual patient can be obtained from Table 3. The cutaneous melanoma samples and the glioblastoma samples (which were needle biopsies) were analyzed by the high-resolution alpha track radiographic technique at MIT. The ocular melanoma samples were analyzed by the prompt-gamma method at BNL.

SUMMARY

The three study objectives stated in the Introduction above have been met to varying degrees.

Objective 1

There was no BPA-related effect on vital signs (monitored for two hours post-administration) or on blood clinical chemistries at all dose levels.

Objective 2

Quantitative data was obtained regarding a) the boron levels in the blood; b) the rate of boron excretion in the urine; and c) the total amount of boron excreted in the urine over a 48 hour period. Using these data, the pharmacokinetics of BPA absorption and excretion have been successfully modeled using a one compartment model with first order absorption and first order elimination. The apparent half-times for absorption into, and clearance from, the volume of distribution were 1.5 and 12.4 hours, respectively. The half-time for appearance of boron in the urine was approximately 9 hours. The fraction of the administered dose of boron that was recovered in the urine within a 48 hour period, ranged from 34 to 65%.

Objective 3

Boron concentrations in tumor samples following BPA doses below 90 mg/kg were below the detection limit of the BNL boron analysis facility. Reliable values for boron concentrations in tumor were obtained for the last seven patients studied (2 glioblastoma, 2 ocular melanoma and 3 cutaneous mela-

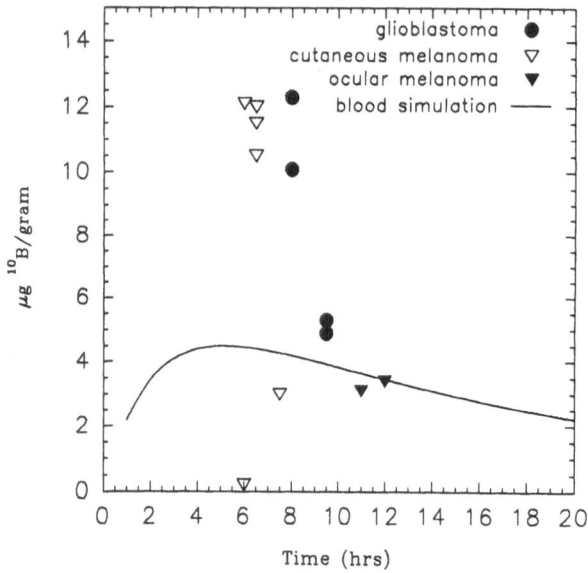

Figure 3. ^{10}B *concentration (μg $^{10}B/g$) in tumor as a function of time after BPA administration. The solid line is a simulation of the ^{10}B concentration in blood at the 189 mg BPA/kg dose level obtained by plotting the solution to the one compartment oral absorption model using the mean values for the pharmacokinetic parameters (see Table 2).*

Table 3

^{10}B Concentrations in Tumor and Blood

Patient	Tumor Type	Dose	Time[a]	^{10}B in Tumor	^{10}B in Blood[b]
		(mg/kg)	(hrs)	($\mu g/g$)	($\mu g/g$)
VZ1	ocular melanoma	27	12	ND[c]	<1.0
BW2	ocular melanoma	54	18	ND	<1.0
LR3	ocular melanoma	54	16	ND	1.2
MS4	ocular melanoma	54	17	ND	<1.0
ES5	ocular melanoma	90	15	ND	1.4
LF6	cutaneous melanoma	90	6	0.2	2.1
JR7	glioblastoma (III/IV)	90	8	10.1;12.3	1.6
JM8	glioblastoma	135	9.5	4.9;5.3	1.2
ES9	cutaneous melanoma	135	6	12.1	2.2
RB10	melanoma (axillary node metas.)	135	6.5	10.5;11.5;12.0	1.8
SZ11	ocular melanoma	189	12	3.4	2.5
JB12	ocular melanoma	189	11	3.1	3.8
SN13	cutaneous melanoma	189	7.5	3.0	2.3

[a] Elapsed time between BPA administration and surgery.

[b] ^{10}B concentration in blood at the time of surgery.

[c] ND, not detected.

noma). The data points were widely scattered and too few to enable any pre-diction of the time course of boron accumulation in the tumor following BPA administration. Some very encouraging boron concentrations in tumor and tumor-to-blood boron concentration ratios were obtained for individual patients (glioblastoma: 11.2 ppm, tumor/blood ratio = 7:1; cutaneous melanoma: 11.3 ppm, tumor/blood ratio = 6:1).

CONCLUSIONS

1) Oral BPA is safe in humans at the 189 mg/kg level.

2) Additional biodistribution studies are needed to determine the time course of the tumor uptake and clearance of BPA for each tumor type.

3) Higher single doses or multiple doses will be required to obtain therapeutically useful levels of boron in the tumor.

ACKNOWLEDGEMENTS

The author thanks Peggy Micca for invaluable assistance including the prompt-gamma boron analyses performed at BNL. This work was supported in part by a grant from the National Cancer Institute (CA42446) to J.A.C., and by the Office of Health and Environmental Research of the U.S. Department of Energy under contract DE-AC02-76CH00016. Accordingly, the U.S. government retains a nonexclusive, royalty-free license to publish or reproduce the published form of this contribution, or allow others to do so, for U.S. Government purposes.

REFERENCES

1. Coderre, J.A.; Glass, J.D.; Fairchild, R.G.; Roy, U.; Cohen, S.; Fand, I. Selective targeting of boronophenylalanine to melanoma for neutron capture therapy. *Cancer Res.* 47, 6377-6383 (1987).

2. Coderre, J.A.; Kalef-Ezra, J.A.; Fairchild, R.G.; Micca, P.L.; Reinstein, L.E.; Glass, J.D. Boron neutron capture therapy of a murine melanoma. *Cancer Res.* 48, 6313-6316 (1988).

3. Coderre, J.A.; Glass, J.D.; Fairchild, R.G.; Micca, P.L.; Fand, I.; Joel, D.D. Affinity of the melanin precursor analog p-boronophenylalanine for tumors other than melanoma. *Cancer Res.* 50, 138-141 (1990).

4. Snyder, H.R.; Reedy, A.H.; Lennarz, W. Synthesis of aromatic boronic acids. Aldehydo boronic acids and a boronic acid analog of tyrosine. *J. Am. Chem. Soc.* 80, 835-838 (1958).

5. Roberts, D.C.; Suda, K.; Samanen, J.; Kemp, D.S. Pluripotential amino acids I. (L)-p-dihydroxyborylphenylalanine (L-Bph) as a precursor of the L-Phe and L-Tyr containing peptides; specific tritiation of L-Phe containing peptides as a final step in synthesis. *Tetrahedron Lett.* 21, 3435-3438 (1980).

6. Fairchild, R.G.; Gabel, D.; Laster, B.H.; Greenberg, D.; Kiszenick, W.; Micca, P. Microanalytical techniques for boron analysis using the $^{10}B(n,\alpha)^{7}Li$ reaction. *Med. Phys.* 13, 50-56 (1986).

7. Solares, G.; Zamenhof, R.; Saris, S.; Wazer, D.; Kerley, S.; Joyce, M.; Madoc-Jones, H.; Adelman, L.; Harling, O. Biodistribution and pharmacokinetics of p-boronophenylalanine in C57BL/6 mice with GL261 intracerebral tumors, and survival following neutron capture therapy. In: *Progress in Neutron Capture Therapy for Cancer*. Allen, B.J.; Moore, D.E.; Harrington, B.V., eds., Plenum Press, New York, 475-478 (1992).

8. Coderre, J.A.; Slatkin, D.N.; Micca, P.L.; Ciallella, J.R. Boron neutron capture therapy of a murine melanoma with p-boronophenylalanine: Dose response analysis using a morbidity index. *Radiat. Res.* 128, 177-185 (1991).

9.   Coderre, J.A.; Joel, D.D.; Micca, P.L.; Nawrocky, M.M.; Slatkin, D.N.
     Control of intracerebral gliosarcomas in rats by boron neutron cap-
     ture therapy with p-boronophenylalanine. *Radiat. Res.* 129, 290-296
     (1992).

RBE IN NORMAL TISSUE STUDIES

Reinhard A. Gahbauer[a], Ralph G. Fairchild[b],
Joseph H. Goodman[c], and Thomas E. Blue[d]

[a]Division of Radiation Oncology
[c]Division of Neurosurgery
[d]Division of Nuclear Engineering
The Ohio State University
Columbus, OH 43210, USA

[b]Medical Department
Brookhaven National Laboratory
Upton, NY 11973, USA

INTRODUCTION

Single dose RBEs for the radiations encountered in BNCT have been determined[1] and can provide useful guidance for the tolerance to be expected. The difficulty of microdosimetric determination of dose[2], the rapidly varying mix of constituent high and low LET radiations with depth, the possible synergistic interaction between high and low LET radiations[3], and the strong dependence of biological efficacy on the distribution of boron within the cell are all factors limiting the usefulness of RBEs. Large animal studies are necessary to determine late effect tolerance. To design, interpret, compare, and apply these studies to clinical use, it has been proposed to divide factors influencing "RBE" into two groups[4]: 1) estimated tolerance dose (ETD), which can be inferred from other experience reasonably well; and 2) compound factor (CF), which is much more variable as a function of microdosimetry and boron compound.

Determination of Estimated Tolerance Dose (ETD)

Tolerance restrictions from known clinical information about tolerance to high and low LET radiations are used. These are somewhat imprecise, but their numerical uncertainty is small compared to the numerical variation of the compound factor. They can further be verified or adjusted as shown below.

For reasons discussed extensively elsewhere[5], we anticipate that BNCT with epithermal neutron beams will be used with a fractionation scheme employing a minimum of 4 fractions. Therefore, the first objective is to establish the tolerance of normal brain to 4 fractions of low LET radiation. Pezner and Archambeau have used a modified version of the Ellis-NSD formula to more realistically reflect brain tolerance[6]. To demonstrate the principle, we use this simple formula, although other formulas can be used.

Formula 1: $TD = BTU \cdot N^k \cdot T^l$

BTU	= Brain Tolerance Unit	N	= Number of Fractions
T	= Time in Days	k	= 0.45
l	= 0.03	TD	= Tumor Dose

In our discussion, we evaluate the formula for 4 fractions and a treatment time of 5 days, using a BTU of 1200 which would reflect the upper limit of acceptable risk to normal tissue. Evaluation of this formula suggests a maximum tolerated low LET dose of 2300 cGy in 4 fractions in 5 days.

The second objective, to estimate the tolerance of normal brain to high LET radiations if only high LET radiations were used, is more difficult because the experience is more limited and because varying fractionations and beams with different RBEs have been used. The influence of fractionation on iso-effect is minimal if more than 4 fractions are used[7]. From clinical information drawn from instances in which more than 4 fractions were used, one may infer that the tolerance of normal CNS is less than 1000 cGy of high LET radiations[8,9]. We apply this limit to any fractionation scheme with more than 4 fractions of only high LET radiation.

Using the 2 restrictions given above, the following formula represents the ETD.

Formula 2: $ETD = D_\gamma + D_h \cdot 2.3 = 2300$

Here D_γ = total dose in cGy for low LET components and D_h = total high LET or particle dose, including components from the $^{14}N(n,p)^{14}C$ and $^{10}B(n,\alpha)^{7}Li$ reactions and fast neutron dose. The dose-limiting factor, 2.3 [2.3=TD/1000], is a factor chosen to restrict normal tissue dose to levels of high LET radiations found acceptable in fast neutron therapy and can be viewed as taking the function of an RBE. In other words, in this case the estimated tolerance of normal tissue was used to determine RBEs rather than the reverse. The measured, single dose RBE of fast neutrons encountered in BNCT, as well as from the $^{14}N(n,p)^{14}C$ reaction, is around 2, while that from the $^{10}B(n,\alpha)^{7}Li$ reaction was found to be 2.3[1,10]. Increasing the number of fractions would effectively increase our factor (just as RBEs would) as the

ETD increases (for 25 fractions ETD = 5600; 5600 = $D_\gamma + D_h \cdot 5.6$). [From Formulas 1 and 2: The factor 5.6 limits D_h to 1000 cGy, as the RBE would.]

Determination of Compound Factor

In large animal, normal tissue studies, it is then possible to derive the compound factor with two sets of experiments (dose escalation studies) comparing the same normal tissue endpoints expressed as a function of ETD.

1) Animals treated without boron:

$$D_{1\gamma} + (D_{1N} + D_{1H}) \cdot 2.3 = ETD$$

This experiment also serves to verify or adjust ETD.

2) Animals treated with boron (only one boron tissue level needed):

$$D_{2\gamma} + (D_{2N} + D_{2H} + CF \ \alpha \ D_B) \ \alpha \ 2.3 = ETD$$

It then follows:

$$D_{1\gamma} + (D_{1N} + D_{1H}) \cdot 2.3 = D_{2\gamma} + (D_{2N} + D_{2H}) \cdot 2.3 + CF \cdot D_B \cdot 2.3$$

Formula 3: $\qquad CF = \dfrac{D_{1\gamma} \cdot D_{2\gamma}}{D_B \cdot 2.3} + \dfrac{D_{1N} + D_{1H} \cdot D_{2N} \cdot D_{2H}}{D_B}$

where
$\qquad D_N$ = Dose from $^{14}N(n,p)^{14}C$
$\qquad D_H$ = Fast neutron dose
$\qquad D_B$ = Dose from $^{10}B(n,\alpha)^7Li$

MATERIALS AND METHODS

If the Compound Factor is determined as described above, Formula 2 becomes:

Formula 4: $\qquad 2300 = D_\gamma + (D_N + D_H + CF \cdot D_B) \cdot 2.3$

where
$\qquad 2300$ = ETD for CNS, 4 fractions
$\qquad D_H$ = Fast neutron dose
$\qquad D_B$ = Boron dose

Since the constituent radiations vary dramatically with depth, Formula 4 has to be determined for a point below to the surface, i.e. $d_{max} \approx 3$ cm, and, in

Table 1

*Dose Rates of the Epithermal Neutron Beam of the Brookhaven
Medical Research Reactor in a Cylindrical Phantom (Data from [10])*

Depth in Phantom	Dose Rate (cGy/min)			
	0 cm	3 cm	8 cm	POP 8+8
Gamma	8.50	16.00	11.00	22.00
Fast neutrons	5.30	0.63	0.00	0.00
$^{14}N(n,p)^{14}C$	0.31	1.25	0.50	1.00
Boron 1 ppm	0.22	0.87	0.29	0.58

POP: Midline in a 16 cm cylindrical phantom irradiated with parallel opposed fields

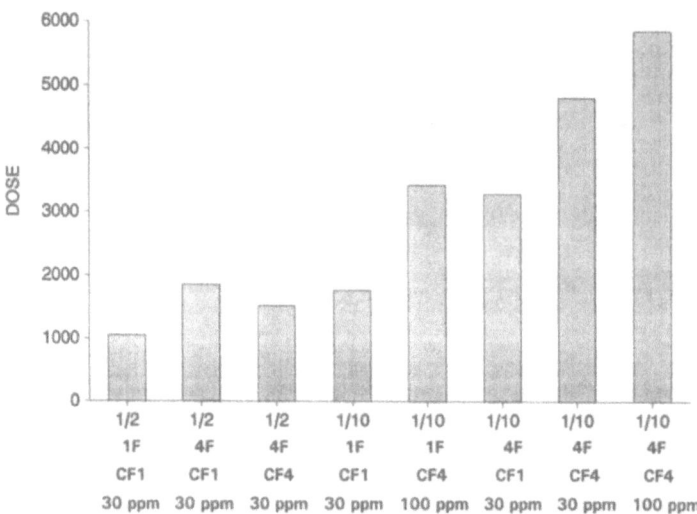

Figure 1. *Maximum midline tumor dose obtainable. Ratio = tumor-normal
tissue boron concentration ratio. Loading = boron concentration in tumor.*

the case in which parallel opposed fields are used, d_{max} at midline. It is then convenient to determine from dose rate tables (Table 1) the depth at which ETD is reached in the shortest treatment time. That time, representing the tolerance limit, is used to determine what minimum TD would be obtained, e.g., at 8 cm depth or at midline. Only physical doses are used, except for the boron dose (which is multiplied by the Compound Factor).

This approach may be useful to arrive at clinically useful tolerance limits and to model the influence on the expected therapeutic gain (midline minimum TD) of parameters such as fractionation T/N tissue ratios, CF, and tumor loading. Examples are shown in Fig. 1.

DISCUSSION

The biological efficacy of boron is known to be a strong function of intracellular distribution, which is generally unknown or imprecisely known for the compounds being investigated for possible use in BNCT. Problems associated with the definition of dose and the use of RBE in BNCT are discussed elsewhere[2]. To account for unknowns inherent to BNCT, Gabel[2] has proposed a compound factor.

Our formula attempts to relate known information on tolerance to the experiment in BNCT and derives the value of the compound factor with only two sets of experiments. This factor is unique to a given boron compound and epithermal neutron beam combination. For compound development, ETD may help to estimate the therapeutic range. It sets a target range for dose escalation studies, which then will define its value more precisely, as proposed above. If the compound factor is not determined, experiments at one boron level would not predict tolerance at another.

In the light of the complexity of the interrelationship of the many physical and biological parameters in BNCT, the approach suggested attempts to arrive at the clinically critical value of tolerance dose with the least number of animal studies. Further, by using one consistent method of tolerance evaluation, it is possible to model the influence of any parameters on the critical minimum obtainable tumor dose. The numbers used in this discussion must be verified by experiments: however, the direction of change should be predictable.

REFERENCES

1. Gabel, D.; Fairchild, R.G.; Larsson, B.; Börner, H.G. The relative biological effectiveness in V79 Chinese hamster cells of the neutron capture reactions in boron and nitrogen. *Radiat. Res.* 98, 307-316 (1984).
2. Gabel, D. Approach to boron neutron capture therapy in Europe: Goals of a european collaboration on boron neutron capture therapy. In: *Proc.*

of the 2nd Eur. Part. Acc. Conf., Nice. Editions Frontières, Gif-sur-Yvette, 283-285 (1990).

3. Ngo, F.Q.; Blakely, E.A.; Tobias, C.A. Sequential exposures of mammalian cells to low- and high-LET radiations. I. Lethal effects following x-ray and neon-ion irradiation. *Radiat. Res.* 87, 59-78 (1981).

4. Gahbauer, R.A.; Fairchild, R.G.; Goodman, J.H.; Blue, T.H. Can relative biological effectiveness be used for treatment planning in boron neutron capture therapy? In: *Tumor Response Monitoring and Treatment Planning.* Breit, A., ed., Springer, Berlin, in the press (1992).

5. Fairchild, R.G.; Bond, V.P.; Woodhead, A.D. *Clinical Aspects of Neutron Capture Therapy.* Plenum Press, New York (1989).

6. Pezner, R.; Archambeau, J. Brain tolerance unit: A method to estimate risk of radiation brain injury for various dose schedules. *Int. J. Radiat. Oncol. Biol. Phys.* 7, 397-402 (1981).

7. Fowler, J.F. Nuclear Particles in Cancer Treatment. Adam Hilger, Bristol, England (1981).

8. Catterall, M.; Bloom, H.J.G.; Ash, D.V.; Walsh, L.; Richardson, A.; Uttley, D.; Gowing, N.F.C.; Lewis, P.; Chaucer, B. Fast neutrons compared with megavoltage x-rays in the treatment of patients with supratentorial glioblastoma: A controlled pilot study. *Int. J. Radiat. Oncol. Biol. Phys.* 6, 261-266 (1980).

9. Laramore, G.E.; Diener-West, M.; Griffin, T.W.; Nelson, J.S.; Griem, M.L.; Thomas, F.J.; Hendrickson, F.R.; Griffen, B.R.; Myrianthopoulos, L.C.; Saxton, J. Randomized neutron dose searching study for malignant gliomas of the brain: Results of an RTOG study. *Int. J. Radiat. Oncol. Biol. Phys.* 14, 1093-1102 (1988).

10. Fairchild, R.; Kalef-Ezra, J.; Saraf, S. Installation and testing of an optimized epithermal neutron beam at the Brookhaven Medical Research Reactor. In: *Neutron Beam Design, Development, and Performance for Neutron Capture Therapy.* Harling, O.K.; Bernard, J.A.; Zamenhof, R.G., eds., Plenum Press, New York, 185-199 (1990).

TREATMENT PLANNING AND OPTIMIZATION FOR

PION THERAPY

Hans Blattmann

PSI, Paul Scherrer Institute
CH-5232 Villigen-PSI, Switzerland

INTRODUCTION

The goal of radiotherapy is the deposition of a high dose in the target volume to inactivate every single cell of a tumor without irradiating the normal tissues to a level giving rise to complications. The negative pions as charged particles have favorable properties concerning physical dose distribution. In addition, at the end of the range they undergo a nuclear reaction with target nuclei, liberating neutral and charged secondary particles, some of them with an increased radiobiological efficiency. For these reasons they were considered particularly suited for radiotherapy. To make optimum use of the particles, a new delivery technique was developed to fit the dose distributions in three dimensions to the target volume. With the developed therapy modality it was expected, after a stepwise optimization of the treatment, to reach an increased local control rate with equal or lower complication probability.

PHYSICAL CHARACTERISTICS OF THE RADIATION

Pions are produced by collisions of high energy protons with matter. Only the negative pion with a mass of 139.6 MeV, 270 times heavier than electrons and with a short half life time of 26 nanoseconds, is of importance for radiotherapy. As a charged particle, the pion has a fairly well defined range and at the end of its track, due to its negative charge, gets captured by a nucleus of the target material. The target nucleus disintegrates and emits on the average three neutrons and some heavy charged particles, all of which have a rather limited range[1]. The heavier of these particles deposit their energy in tracks of high linear energy transfer (LET). These particles have an increased radiobiological efficiency, particularly for anoxic cells which are less sensitive for low LET radiations by a factor of nearly three compared to normal cells. This fact was the reason for the attention that the pions received initially as radiotherapeutic agent, but is also the reason why pions had to be carefully studied before the first application on humans.

To achieve reasonable dose rates, high currents of protons, or pion beam channels with a large collection solid angle, are necessary. At PSI a proton beam of up to 20 μA is split off the main proton beam and guided onto a pion production target for medical applications.

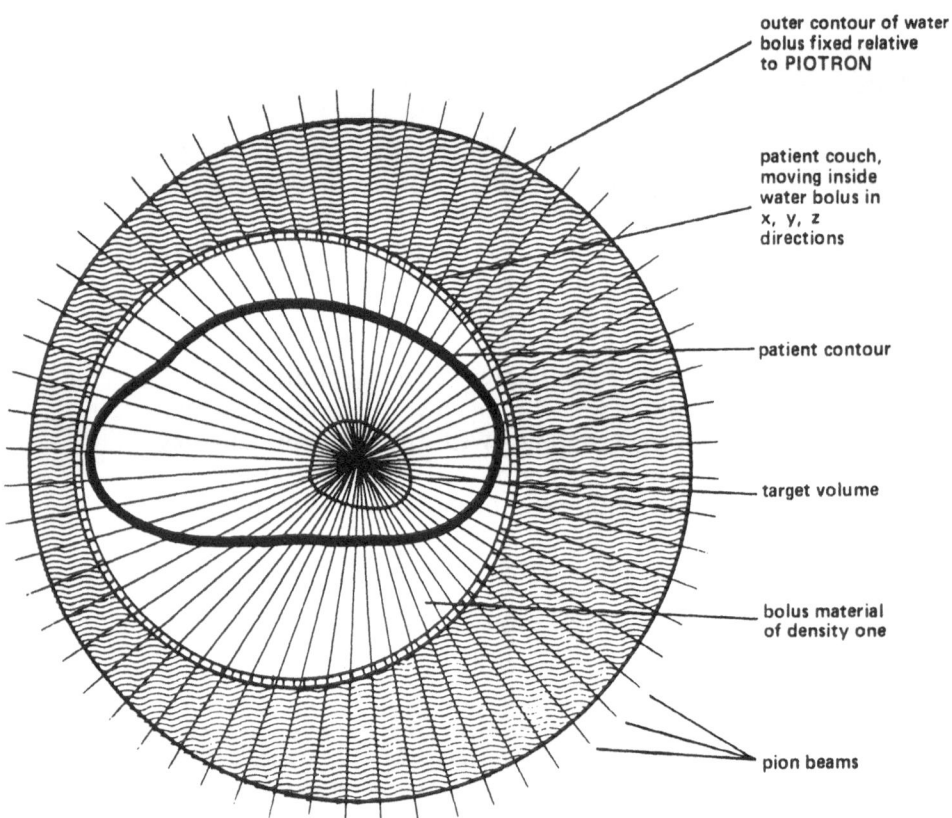

outer contour of water
bolus fixed relative
to PIOTRON

patient couch,
moving inside
water bolus in
x, y, z
directions

patient contour

target volume

bolus material
of density one

pion beams

Figure 1. *Spot scan. The patient is positioned on a couch inside a water bolus, coaxial with the Piotron. By scanning the patient in three dimensions the irregularly shaped target volume is irradiated by a spot produced by 60 pion beams.*

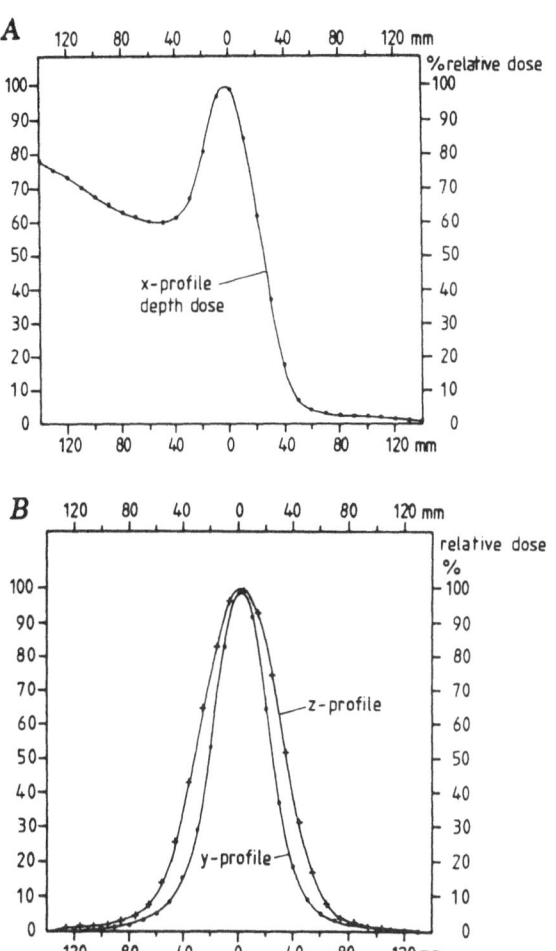

Figure 2. *Depth dose profile for a single beam (A) and dose profiles for a 60 pion beam spot (B) on the axis of a 62.8cm diameter water phantom.*

PIOTRON

The 60 beam pion applicator or Piotron is an improved version of the Stanford Medical Pion Generator (SMPG)[2,3,4]. Unlike the plans of Stanford, protons of 590 MeV are used to produce pions. The protons hit the beryllium production target, where they lose approx. 21 MeV and are stopped in a copper absorber immediately behind the target. A shielding wall of 4 m of iron protects the patient from unwanted radiation from the target and the beam stop. Even though different target lengths and materials are available, most patients have been irradiated using a 7 cm beryllium production target, 6 mm in diameter, slightly larger than the proton beam diameter. Pions and other secondary particles originate in this target with a high energy and angular distribution. Pions of the appropriate energy are selected by a first ring of 60 superconducting bending magnets. The 60 pion beams are bent a second time, entering the patient treatment chamber in a vertical plane. Each beam can be switched on and off by a copper slit system, situated between the two magnet rings. The patient is positioned in the beam lying in a couch with a customized mould. The patient transport system, responsible for accurate positioning and movement of the couch during treatment inside the Piotron, guarantees a precision for the relative movements of the platform on which the patient couch is mounted of better than 0.5 mm.

Spot - Scan - Application - Technique

The spot scan (Fig. 1) was the result of the attempt to get, in the first approximation, a homogeneous biological dose distribution. If the pions are stopped on the axis of a cylindrical phantom, which is mounted coaxial to the Piotron, the dose contributions of the individual particles, i.e. pions and contaminating electrons and muons are proportional to their abundance, multiplied with the corresponding energy loss. If this spot is scanned across the target volume, each volume element receives the same mixture of particles. The dose contribution for a beryllium target due to electrons and muons is less then 15%. Depth dose profiles for a single pion beam and spot profiles along the Piotron axis (Z-profile) and perpendicular to the Piotron axis in a 62.8 cm diameter water cylinder, are given in Fig. 2. To keep the dose spot invariant during scanning, the patient is positioned for practical considerations inside a water cylinder resulting in a homogeneous stopping power distribution throughout patient and bolus. The spot scan technique can be performed with 60 or less beams, the number of beams is generally constant during the entire treatment for a patient. A cylindrical bolus of solid, water-equivalent material is produced, and the patient positioned in a half cylindrical couch supplemented to a cylinder in the treatment region.

Conformation Radiotherapy

The aim of the development of conformation therapy is to deliver the prescribed dose to the target volume and sparing normal tissue outside as completely as possible.

The geometrical properties of the Piotron, i.e. 60 beams in a vertical plane focused onto the axis of the Piotron, and the physical properties of the charged particles, made it necessary to develop a dynamic application technique to treat larger target volumes. As the 60 coils are electrically connected in series, all of the 60 beams have the same range. If a circular cylinder positioned coaxial with the Piotron is irradiated with pions of the appropriate range to let them stop on the axis, a dose spot is formed. In the vertical plane the diameter of the spot is dependent on the multiple scattering of the pions in the material in front of the stopping volume. For the spot scan technique the spot is scanned throughout the target volume plane by plane, resulting finally in a three dimensional conformation of the treatment volume to the target volume.

Treatment Planning

Treatment planning is based on CT scans taken in steps of 5 to 10 mm over the entire target volume with safety margins on either side. To guarantee the same geometry for treatment planning and irradiation, the patient is scanned in his couch prepared for the treatment. For CT-imaging the couch is used with the foam substituted by tissue-equivalent material in the treatment region, as this increases the diagnostic quality of the pictures. In each of the CT-slices the contour of the target volume is entered on the computer screen by track ball. The desired dose distribution then consists of a homogeneous dose of 100% inside this target volume and a rapid fall-off outside.

This calculation is done using the *physical* dose for those cases where all the 60 beams are used. A symmetrical dose distribution of the spot can be assumed in this case. For treatment plans where only part of the beams are used, either because the lesion is eccentric or because one would like to avoid a specific area, the *biological* dose is optimized. The spot dose distribution is determined by superposition of the dose distributions for the single beams used for the specific case. The biological dose takes into account the changing radiation quality along a single pion beam, which results from the varying contributions of primary and secondary particles. In the treatment planning program[5] hypothetical dose spots are deposited on a regular grid. For each spot the dose necessary to achieve the desired dose distribution is calculated. For the treatment, the dose per spot is transferred into a scan speed of the spot with a given dose rate. The calculation of the biological dose takes two high LET dose components, the star dose (ALU) and the neutron dose (NEU), into account.

Physical Dose Determination

Precise dosimetry is a prerequisite of an efficient radiation therapy. For a complex radiation field, as in pion therapy, dosimetry is even more demanding. For the Piotron, as a first step, the phase space of the single beams have been measured. In the entrance region pions, electrons and muons but also secondary particles from pion stars in flight, contribute to the dose. In the stop region of the pions, the importance of secondary particles relative to the primary particles increases. The number and energy spectrum encountered depends to a large extent on the chemical composition of the target material[6].

Detectors

Dosimetry for pions is done primarily with tissue equivalent ionization chambers in water and polyethylene phantoms. For specific measurements TLD (thermoluminescence dosimeters), diodes and activation of aluminium are used. Microdosimetric measurements have been performed in single beams as basic information on radiation quality.

Treatment Plan Verifications

The dynamic application technique with individual treatment plans for each patient asks for a verification of the dose distribution at least for some positions. After some initial tests in a specially built anthropomorphic phantom, it was decided that for routine verification a modular, homogeneous polyethylene phantom is better suited. The accumulated results for a group of patients have been compared with the calculated doses for the corresponding positions (Fig. 3a).

In Vivo Dosimetry

In vivo dosimetry is performed during one of the first fractions for every patient. If possible, catheters with ionization chambers are used in the target volume. To verify the treatment planning calculations in vivo, the bladder irradiations were of special interest at the beginning of the clinical program. These measurements answered directly the question whether or not it is permitted to neglect density inhomogeneities for the calculation, even if rela-

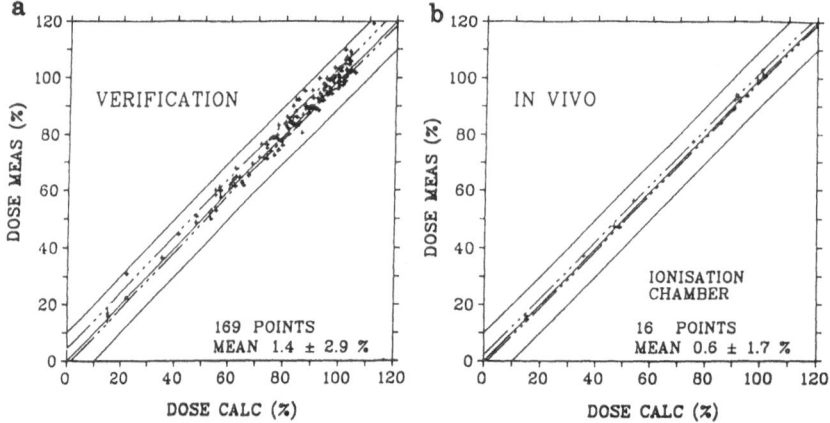

Figure 3. *Comparison between calculated and measured dose for phantom measurements (verification) and ion chamber measurements in patients (in vivo).*

tively thick bones, as in the pelvis, are passed by the pions. A good correspondence was seen between calculated and measured doses in vivo (Fig. 3b).

Biological Dose Determination

The treatment planning program is based on a simplified model describing biological efficiency based on data originating from microdosimetric measurements. The model has been calibrated by results of biological experiments with Ehrlich ascites tumor cells irradiated with a dose corresponding to a single fraction of a radiotherapeutic treatment scheme. Another series of biological experiments have been performed to check calculated biological doses. Several parameters have been systematically varied and their influence checked on several biological endpoints.

The physical dose distributions of a single beam has been measured with an ionization chamber and used as a basic input into the treatment planning program, it was named (ION). To take into account the higher efficiency of the pions at the end of their range, as well as the charged particles emitted from the pion star, the star dose (ALU), determined by aluminium activation, was measured. And finally the neutron dose (NEU) originating from neutrons emitted from the star, but which is not deposited locally, is taken into account by a neutron dose model, based on data of Scillaci and Roeder[7] who have calculated neutron dose distributions of a pion stop by use of the evaporation model and Monte Carlo calculations. Additional input came from microdosimetric measurements[8]. The star dose and the neutron dose contribution together form the high LET contribution of the pion dose. The total of these contributions have to be in accordance with the microdosimetrically determined high LET dose in the pion spot. Of the 35% high LET dose contribution (LET > 75 keV/μm) measured by Schuhmacher and Menzel, 15% has been attributed to the neutron dose and 20% to the star dose. The RBE related to the two high LET contributions has been assumed to be the same, and the value has been chosen to yield an overall RBE of 1.5 for a 16 - 20 fraction treatment of a 300 cm^3 volume. It may seem surprising, that the RBE for pion stars and neutrons should be the same, but the neutron spectrum with energies of up to 100 MeV is expected to have a lower RBE than that of lower energy neutrons, and the star dose only contains the high LET contribution of the dose of a pion star. With these assumptions the biological dose is

$$BIO_{2Gy} = ION \cdot (ALU + NEU) + RBE_{2Gy}(ALU + NEU)$$

$$RBE_{2Gy} = 2.13$$

This equation is valid for 2 Gy and has to be modified for doses outside the target volume, which are below 2Gy.

$$BIO = BIO_{2Gy} \times F_d$$

where F_d is 1 for 2 Gy and increases for lower doses. F_d is calculated from results of cell experiments[9].

Biological Dosimetry Experiments

The primary biological test system for dosimetry was growth of single cells exposed under various conditions in liquid or in gelatine.

Of special interest was also the determination of the effect on a tissue of an animal, irradiated with a therapeutically significant dose. As one of the questions to be investigated was the dependence of the RBE from treatment volume size, and as the aim was to use a small animal, it was necessary to use a test which could be done with whole body irradiated animals. The test used was the determination of the survival of intestinal crypt cells of mice after one or four doses[10]. With this test system the RBE for therapy conditions for the three centers at Los Alamos, Vancouver and Villigen have been compared. No significant differences have been found for fractionated irradiation.

Radiobiological Results

A large variety of radiobiological experiments has been performed primarily to answer questions in connection with therapy. Of basic interest was the measurement of the RBE in the plateau and peak region for different biological test systems. As in the therapy situation there is always a mixture of plateau and peak dose contributions, i.e. low LET and high LET, the RBE has also been determined for different sizes of irradiation volumes[9].

Asymmetric beam geometries were of special interest, as for these cases the biological dose has been optimized instead of the physical dose. The validity of the calculations has been verified by biological experiments. The high LET contribution increases from the proximal border to the distal surface of the treatment volume from 33 to 40%, resulting in a 7% increase in RBE according to the treatment planning program.

Another area of interest was the dose delivery pattern in time. Due to the spot scanning technique some volume elements received their main dose right at the beginning by the passage of the spot. For the rest of the irradiation time, they accumulated little dose from passing pions with low LET. For other volume elements the sequence of the dose depositions was opposite. It was therefore an open question if due to repair during treatment time the irradiation was less efficient for one area compared to another one. Cell culture experiments demonstrated differences of lower than 4% for treatment times of one hour and even less for shorter irradiations. It was therefore considered unnecessary to correct for these differences in clinical practice.

CLINICAL PROGRAM

International Advisory Board

An international advisory board was formed for the selection of the sites, fractionations, dose, dose per fraction, RBE etc. The implementation of pion therapy was planned in three phases. The first phase was to compare the effects of pions and x-rays for human cancers metastatic to the skin or subcutaneous sites and on the overlying skin. The second phase would be treatment of primary advanced cancer, and the third, randomized clinical trials for selected indications. The advisory group was also considered to be the basis for patient referral for this program especially from foreign countries.

Selection of Indications

For phase 2 of this new modality, indications were chosen which were considered to yield non optimal results with conventional treatment. The selection included bladder, pancreas and glioblastoma. For the pion therapy facility, of primary importance was the treatment of bladder carcinoma because it was expected that for this treatment, conformation would be important to spare normal tissues adjacent to the bladder, to increase the biological efficiency in the target volume by the high LET component of the irradiation and last but not least, because it was possible, to verify in vivo the target dose. Pancreas and glioblastoma were primarily chosen because of their unsatisfying results with other treatment techniques, in the hope that the increased RBE of the pions in the tumor tissue would improve the results.

Selection of Dose, Dose per Fraction, RBE

To enable comparison of the results with existing treatment techniques the number of parameters to be varied have to be kept as small as possible. The radiobiological results of neutron irradiations suggested that for radiations with a high LET dose contribution, the number of fractions could be reduced without loss of therapeutic gain, but it was also felt that for the first trials the number of fractions should be the same as for the treatment modality for intercomparison. To keep the number of fractions as low as possible the Manchester technique of 20 fx in 4 weeks was adapted to the schedule of the accelerator at PSI and 20 fx in 5 weeks were chosen for pion therapy.

The selection of doses per fraction was based on experience at Los Alamos, for irradiation of the skin, which was done with slightly less fractions (13 fx).

A modified Ellis formula was applied if doses had to be converted from one fractionation scheme to another one.

Dose Escalation Program

At the start, the dose per fraction for phase 2 was selected, based on conservative estimates of tissue tolerance by the Los Alamos experience and available radiobiological data. A dose 5% less and 5% higher was calculated and alternate patients treated with these schedules. After a three months period, the dose plan was evaluated. If no significant effects were encountered, a second group of patients treated with a 10% higher dose bracket was planned, and so on, until significant acute reactions (grade 2 or higher) were encountered. At this point, no changes should be made until at least 9 months of observation can be carried out in order to assess late effects. The goal of this dose escalation program was to evaluate tumor response as a function of dose, fractionation, and volume and to evaluate normal tissue response and tolerance. Important aspects of treatment planning and technical aspects of delivering the radiation in an attempt to maximize the therapeutic ratio by minimizing the normal tissue volume to be irradiated were expected.

Phase 2 started in 1982, with patients treated for bladder tumors with total doses of 3000 and 3200 cGy pions. A technical failure of the Piotron in the summer of this year interrupted the treatment of some of the patients of this group. Those who received the full treatment had no complication until late fall, and it was therefore decided, even though the number of patients was small, to give the next patients 3400 and 3600 cGy total dose. These doses were unchanged during the following year. Gradually patients in these two dose groups started to develop late complications, shrunken bladder and fibrosis in the tissues outside the bladder. As a consequence the doses were reduced, with indications limited to those patients who had a local invasive tumor in the bladder and who did therefore not need a treatment of the entire bladder wall. This resulted in a substantial drop of bladder patient accrual (Fig. 4). The therapeutic dose determined in the dose escalation program was 33 Gy in 20 fractions in 5 weeks. This treatment scheme was then adopted for all pelvic treatments. For practical purposes, the doses are converted whenever necessary into photon doses by multiplication with a factor of two. This factor is not an RBE but includes also the differences in fraction doses between pion therapy and photon therapy. The corresponding RBE would be 1.7. In addition to the dose reduction, the target volume contours were drawn much tighter to the visible tumor volume. These measures improved the therapeutic results for bladder tumors drastically. The sharp increase of dose effect curves for complications is also obvious from Fig. 5. Important additional information came also from the few patients who received the remaining dose to their target volume following a gap of about four months. Even though the total doses were designed to be well within tolerance according to the Ellis formula with coefficients adjusted for pions, the late reaction was considerably underestimated.

Another program which started early was high grade astrocytoma and glioblastoma treatment. Doses as high as 60 Co-60 Gy equivalent did not

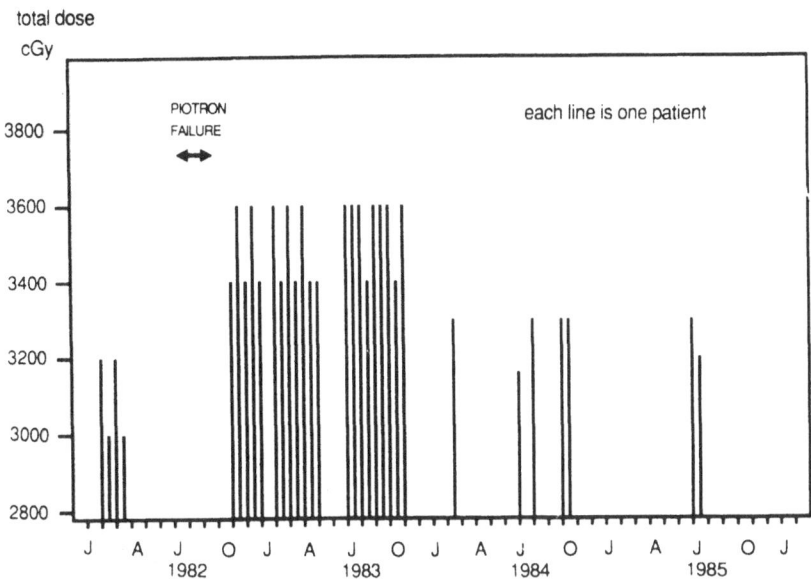

Figure 4. *Patient accrual and total doses for bladder patients in the years 1982 to 1985.*

achieve local control for glioblastoma. An accelerated treatment, reducing the overall treatment time by giving two full doses with a minimal separation of at least 6 hours on some days to 14-17 days, failed to demonstrate a difference in survival time or local control[11].

CLINICAL EXPERIENCE

At the beginning of the pion therapy, the expectations of the clinicians were especially very high. It was generally assumed that the reason for the failure of photon radiotherapy in local control was usually connected with anoxic tumor cells being more radioresistant to low LET radiation, or that they were inherently more resistant. Both groups of cells were expected to be more efficiently treated with high doses of densely ionizing particles. Furthermore the treatment with pions was assumed to protect better normal tissues outside the target volume, due to the favorable dose distribution. Without decreasing radiation burden to the surrounding normal tissues the tumor dose could be increased and also the inactivation probability increased. With accumulating experience it was obvious, that the radiation quality for the normal tissue not only inside but also outside the target volume was the same as that for the tumor. Not until the dose was reduced to a level of approx. 30%

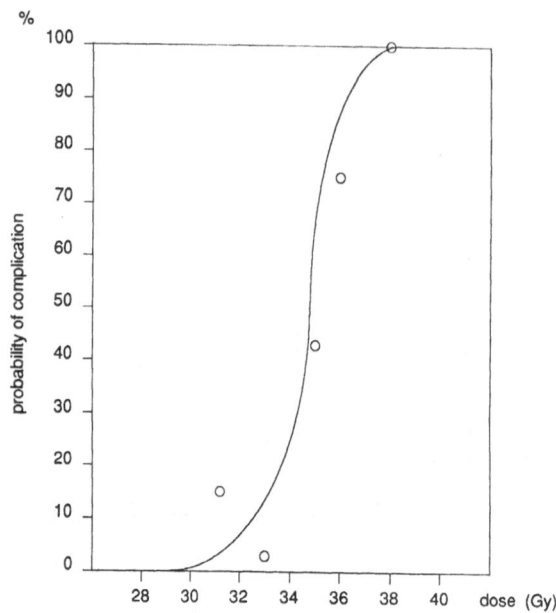

Figure 5. *Probability of complication as a function of dose.*

did the RBE significantly change from the higher value in the treatment volume to a value close to one. This meant that the important normal tissues adjacent to the target volume were not protected by a lower RBE. In other cases, the increase of the biological effect in the target volume was not enough to inactivate the tumor (glioblastoma). For other indications even a local control could not influence the course of the disease, because metastases, not identified before treatment, determined the destiny of the patient. Very encouraging results were found especially for large tumors which could not have been treated with a curative dose with another radiation source. For specific groups, e.g. large retroperitoneal soft tissue and bone sarcoma, conformation radiotherapy with pions could be regarded as the treatment of choice[11,12]. Even though these indications would not justify a clinical pion therapy facility, the effort necessary for these treatments are in a good relation to the result. It is not possible yet to decide if the results are primarily a consequence of the increased RBE of the pions or of the dose conformation, or if the combination of both is necessary.

The present program concentrates on large, inoperable retroperitoneal or pelvic soft tissue and bone sarcoma, bulky cervix carcinoma and prostate cancer.

REFERENCES

1. Blattmann, H. Type and energy spectra of secondaries from interactions of pions, and relevance to radiotherapy. In: *Nuclear and Atomic Data for Radiotherapy and Related Radiobiology*. International Atomic Energy Agency, Vienna (1987).
2. Boyd, D.; Schwettman, H.A.; Simpson, J. A large acceptance pion channel for cancer therapy. *Nucl. Instr. Meth.* 11, 315-331 (1973).
3. Kaplan, H.S.; Schwettman, H.A.; Fairbank, W.M.; Boyd, D.; Bagshaw, M.A. A hospital-based superconducting accelerator facility for negative pi-meson beam radiotherapy. *Radiology* 108, 159-172 (1973).
4. Pistenma, D.A.; Li, C.G.; Bagshaw, M.A. Basic considerations in simulated treatment planning for the Stanford medical pion generator (SMPG). *Int. J. Radiat. Oncol. Biol. Phys.* 2, 345-356 (1977).
5. Pedroni, E.; Blattmann, H.; Salzmann, M.; Walder, E.; Crawford, J.F.; Dietlicher, R.: Cordt, I.; Schäppi, K.; von Essen, C.F.; Perret,C. Treatment planning and dosimetry for the pi-Meson therapy facility at SIN. *Proc. Int. Conf. on Application of Physics to Medicine and Biology*. G. Alberi; Z. Bajzer, P. Baxa, eds., World Scientific, Singapore, 1-25 (1983).
6. Muenchmeyer, G.; Amols, H.I.; Büche, G.; Kluge, W.; Matthäy, H. Moline, A.; Randoll, H. Energy spectra of charged particles emitted following the absorption of negative pions stopped within oxygen-containing organic compounds. *Phys. Med. Biol.* 27, 1131-1149 (1982).
7. Scillaci, M.E.; Roeder, D.L. Dose distribution due to neutrons and photons resulting from negative pion capture in tissue. *Phys. Med. Biol.* 18, 821-829 (1973).
8. Schuhmacher, H.; Menzel, H.G.; Blattmann, H.; Muth, H. Proportional counter dosimetry and microdosimetry for radiotherapy with multiple pion beams. *Radiat. Res.* 101, 177-196 (1985).
9. Pohlit, W.; Jüling, L.; Blattmann, H.; Pedroni, E.; Menzel, H.G.; Schuhmacher, H. Dependence of the RBE on tumor volume in pion treatment of tumors on the "spot scan" mode. *Medical Newsletter SIN* 5, 31 (1983).
10. Raju, M.R. Heavy particle radiotherapy. *Academic Press, New York, London, Sydney, Toronto, San Francisco (1980)*.
11. Greiner, R.; Blattmann, H.; Thum, P.; Bösiger, P.; Coray, A.; Kann, R.; Lahtinen, T.; Reinhardt, H.; von Essen, C.F.; Zimmermann, A. Anaplastic astrocytoma and glioblastoma: pion irradiation with the dynamic conformation technique at the Swiss Institute for Nuclear Research (SIN). *Radiother. Oncol.* 17, 37-46 (1990).
12. Thum, P.; Greiner, R.; Blattmann, H.; Coray, A.; Zimmermann, A. Pionen-Strahlentherapie nichtresektabler Weichteilsarkome am schweizerischen Institut für Nuklearforschung (SIN). *Strahlenther. Onkol.* 164, 714-723 (1988).

DOSE CALCULATIONS BASED ON IMAGE RECONSTRUCTIONS

Floyd J. Wheeler and **Daniel E. Wessol**

INEL BNCT Program
Idaho National Engineering Laboratory
EG&G Idaho, Inc.
Idaho Falls, ID 83415-1575, USA

INTRODUCTION

Before initiation of human clinical trials of Boron Neutron Capture Therapy (BNCT), there must be a significant amount of confidence in dose prediction. While there is a considerable amount of experience in conventional photon irradiations, as well as a growing base in fast-neutron and charged-particle irradiations, procedures and methods are just being developed and no standards exist for BNCT dose predictions.

The BNCT Program at the Idaho National Engineering Laboratory (INEL) has been developing a three-dimensional treatment planning system for BNCT[1]. This system provides a means to perform three-dimensional reconstructions based on patient images to perform precise neutron and gamma transport calculations for the patient model, and to provide any and all output display required for therapy optimization and radiation effect correlations. Radiation transport is currently being performed using the Monte Carlo method. Objects outside of regions defined by the medical images can be added to the model to represent remaining patient features, shielding, and other structure. This system is now being validated[2] and will be used in large animal research programs.

This paper briefly describes the image reconstructions, radiation transport methods, and the comprehensive editing capabilities. A demonstration of the application of the system to a human patient (not an actual BNCT patient) is provided, showing a comparison of two different epithermal-neutron beams (10 cm diameter) and an approach to optimization of the treatment plan.

Boron Neutron Capture Therapy, Edited by D. Gable
and R. Moss, Plenum Press, New York, 1992

METHODS

BNCT planning inherently requires three-dimensional modelling with rigorous solution of the Boltzman transport equations for flux and event rates. Photon therapy, where scattered rays are of secondary importance, can usually be planned with more simple one- and two-dimensional methods. For intermediate- and low-energy neutrons, however, the importance, to absorbed dose, most often increases following a scatter event and one- and two-dimensional methods are entirely inadequate, except for some limited applications. A scheme of the INEL BNCT Program's computational system is provided in Fig. 1. The system consists of compatible software modules for anatomical reconstruction modelling, radiation transport computation, and for dose-pattern display. A large physical data base is incorporated and contains interaction cross sections for neutrons, photons, and charged particles significant to BNCT. Expert files containing complete boundary conditions for various neutron-source facilities, composition specifications, geometric models, etc., are accumulated with experience for use in subsequent applications.

Anatomical Reconstruction Modelling

The three-dimensional geometric model consists of regular-geometry primitives and/or non-uniform, rational B-spline surfaces (NURBs)[3]. With regular-geometry primitives, objects can be modeled as simple or complex combinations of spheres, cubes, cylinders, etc., and irregularly-shaped objects (i.e., anatomical features) can be only approximately modeled. NURBs, however, can represent anatomical features as precisely as desired. NURBs, a powerful free-form curve and surface representation system may be best described as a naturally smooth curve (or surface) that is represented as a piecewise polynomial expression. NURBs are easy to form into complex sculptured shapes and require much less storage than the polygonal representation

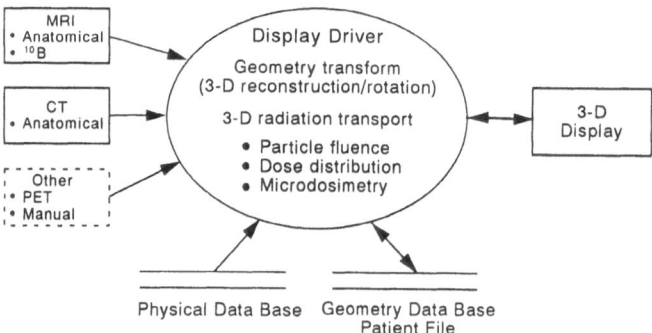

Figure 1. *Simplified treatment planning system.*

that they have replaced. Because of the rational property of NURBs, they can represent exactly conical sections. The curve generated by the use of the B-spline basis is given by, $\alpha(t)$, the position vector (x,y) along a curve as a function of the parameter, t

$$\alpha(t) = \sum_{i=0}^{n} P_i B_{i,k}(t) \tag{1}$$

where P_i are the user-specified control points chosen to fit an object of interest and $B_{i,k}(t)$ are the ith weighting or basis functions of order k (degree +1). Simply stated, the position vector is a weighted sum of the control points at parameter, t.

Once an object has been defined by three or more curves (quadratic fit) oriented in planes along an orthogonal direction, an enclosing surface for that object can be constructed. This surface is described by a tensor product spline, which maps a rectangle into Euclidean three dimensional space and can be regarded as a mathematical extension of the curve to the surface. Just as points control the shape of a B-spline curve, B-spline curves control the shape of the NURB surface.

To utilize this capability, designed for computer graphics applications, in radiation transport analyses, the decision was made to modify the University of Utah's Alpha_1[4] geometry spline-based modelling system so that it would create B-spline geometries suitable for INEL's Monte Carlo transport code. These modifications resulted in a spline-based geometric editor that was appropriately named BNCT_Edit. In addition to interactively carving out three-dimensional geometry from an arbitrary image space (in our case, medical images), BNCT_Edit will also graphically display the results of the Monte Carlo analysis.

Now that free-form geometries are available, we are faced with the issue of how to intersect the spline surfaces to return boundary-intersection distances to the Monte Carlo collision routines. Since a ray tracer was developed at the University of Utah[5] that could compute the surface intersection distances for the free-form geometries, it was a relatively straightforward task to pass this information to the Monte Carlo process analogous to the way the distance to boundary is passed back by the combinatorial geometry routines.

A more detailed treatment of the construction of NURBs surfaces and surface intersections is found in Reference 6. This reference, in turn, provides references to the individual topics.

When reconstructing the anatomical model using medical images, image slices are displayed on the computer screen and used as templates. The outline of a feature is simply and naturally specified by clicking the mouse at border points. The computer fits a smooth outline curve to the points (Fig. 2) and adjustments can be made as desired. These fits can be copied to a new image and the outline curves can be adjusted to fit the feature on that slice. When all image slices are specified, the computer "stitches" all outlines from

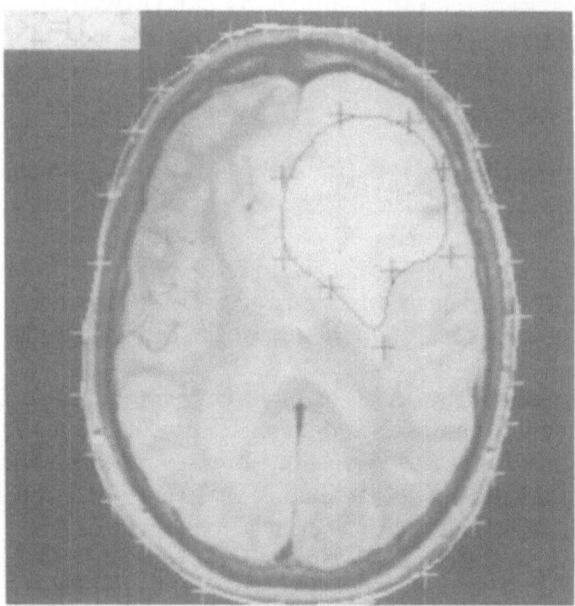

Figure 2. *Screen display of MRI of glioma patient showing B-spline control points and curves representing tumor, edema, and outer surface of skin.*

all slices together to form the three-dimensional model. Fig. 3 shows a wireframe rendition of a partial reconstruction and Fig. 4 shows a rendering of a partial reconstruction. Although not yet a part of the system, automatic surface-fit algorithms from images, such as a modified version of the "marching cube" algorithm or the automatic segmentation algorithm[7], could also be used to assist in the anatomical reconstruction.

Analytical Radiation Transport

Radiation transport is accomplished using the Monte Carlo method. Three-dimensional deterministic methods could also be used. These would require an additional step in which the three-dimensional model is discretized into mesh elements. Conventional Monte Carlo methods, embodied in existing codes, are not adequate for the goals of BNCT dosimetry and improvements have been made. These are, in part: (1) the capability to perform particle transport in NURBs geometry, (2) incorporation of a subelement mesh to conveniently and rapidly obtain intraregional detail in flux and dose, and (3) use of a "converged," induced gamma source for efficiency in the gamma dose calculation. Details of these methods will not be presented here, but will be provided in the proceedings of an upcoming conference[8].

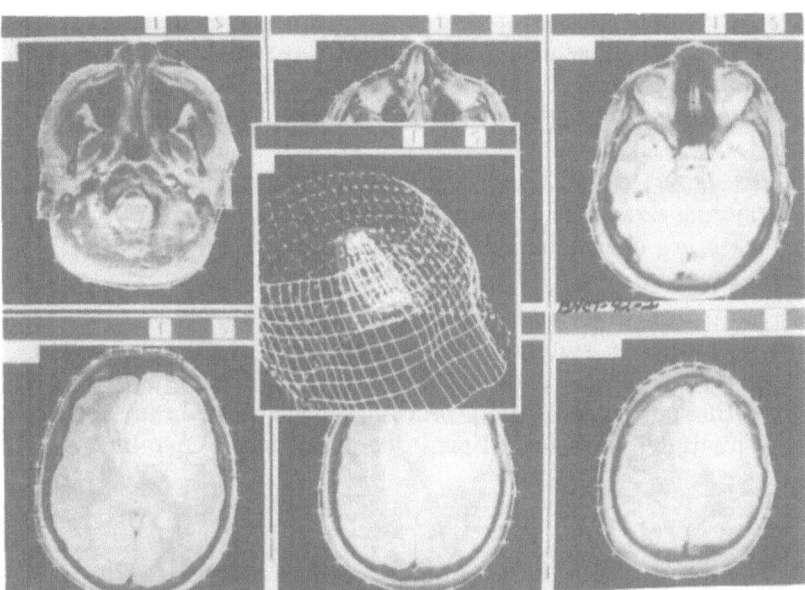

Figure 3. *Wire-frame rendering of partial reconstructions based on several MRI slices.*

Table 1. *Validation 10x10 cm^2 BMRR Beam/10x10 cm^2 Lucite Cube*

Quantity	Discrete Ordinates Results	Monte Carlo Results	
	Reg. Geom.	Reg. Geom.	B-Spline
Maximum thermal-neutron flux (n/cm^2·s·MW)	5.17E+8	5.46E+8 (± 0.03E+8)	5.43E+8 (±0.03E+8)
Maximum gamma dose rate (cGy/MW·min)	1.68	1.79 (±0.01)	1.80 (±0.01)
Maximum ^{10}B dose rate (cGy/MW·min)	0.255	0.240 (±0.002)	0.239 (±0.002)

- Cube is cocentric with beam and incident surface is coplaner with outside of delimiter

- Normalized to beam current of 5.18E+8 (n/cm^2·s)

Validation of the Monte Carlo module is ongoing. A simple comparison was made for a lucite cube irradiated by a 10x10 cm field. The incident radiation field was that calculated for the epithermal-neutron source from the Brookhaven Medical Research Reactor (BMRR), located at Brookhaven National Laboratory, Upton, NY[9]. A NURBs representation of the cube was constructed using eight control points at each of three planes. There were two points at each corner of each plane and, to the eye, there was no difference in the NURBs cube and an actual cube. The "NURBs" solution for peak thermal neutron flux and gamma dose was compared to the solutions from two validated codes. Results are shown in Table 1. Validations have also been reported elsewhere[2].

Results from the radiation transport calculation can be provided in the form of point, line, surface, or volume edits. A point edit is obtained by specifying an x,y,z spatial coordinate obtained from a desired area in a medical image plane or with knowledge of the geometric model. A table is then provided, displaying available data for all components of dose and relative

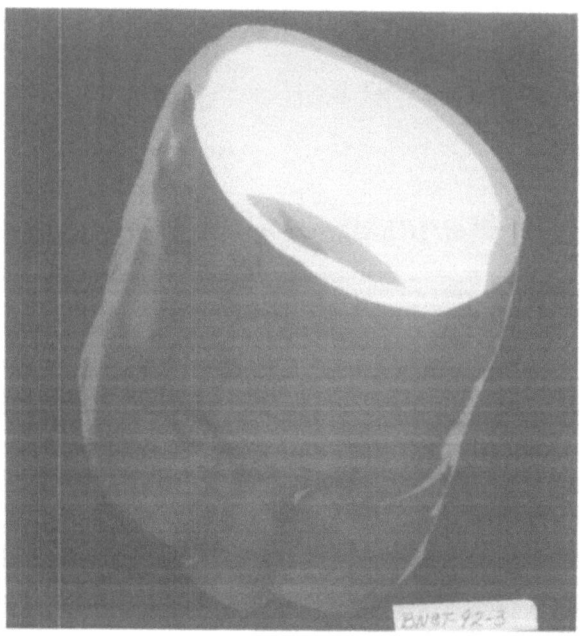

Figure 4. *Surface rencering of partial reconstruction showing tensor-spline surface representation.*

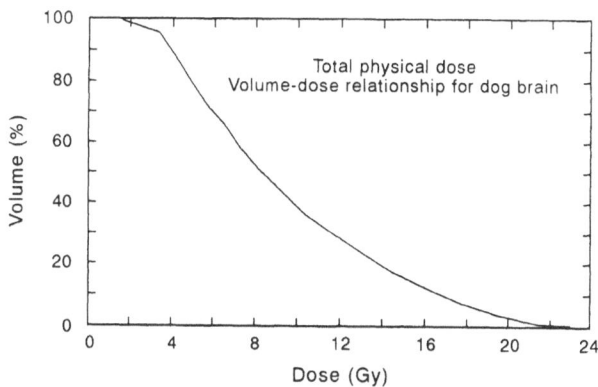

Figure 5. *Volume edit.*

biologic effectiveness (RBE) dose to healthy and tumor tissue, along with microdosimetric data (to the best of present knowledge). A line edit is obtained by specifying only two points. A line plot is then obtained for dose or RBE-dose components. A surface edit is the familiar isoflux or isodose plot and can be obtained for planes perpendicular to any of the three coordinates. Slant planes could also be easily incorporated. A volume edit is one of the powerful and most potentially useful presentation results. An example of a volume edit is provided in Fig. 5. This edit shows the dose volume relationship for total equilibrium blood-brain dose for one of the BMRR dog irradiations. As shown in Fig. 5, almost 100% of the dog brain realizes at least 2 Gy dose and only about 3% realizes a dose >20 Gy.

APPLICATION TO HUMAN BNCT PLANNING

A glioma patient model was constructed and an optimized plan developed to initiate a system exercise. A full optimization study was not developed because this was not a real application, but a trial exercise. The head model was a NURBs reconstruction based on axial image slices from the lower jaw to the top. The lower part of the jaw, the neck, and the upper body were modeled with combinatorial-geometry primitives since these objects were outside the image data.

A 10 cm diameter epithermal neutron beam was used in the calculations. It was assumed that the directionality of the angular flux could be specified and this was assumed to be hemispherically isotropic for one set of calculations. The angular flux was assumed to be monodirectional for a second set of calculations. The neutron energy spectrum was assumed to be similar to the BMRR epithermal neutron beam.

Target tissue was assumed to be simply the tumor itself. In Case A (Fig. 6), the beam aperture was assumed to be centered on the tumor and above the head in an axial orientation. In Case B (Fig. 7), the beam was assumed to be targeted toward the tumor and in an oblique orientation. In the computational system, simulated beam applications such as these are quickly specified assuming the beam characteristics have been previously determined and are stored in an "expert" source file. First, a target point (i.e. center of the tumor) and a beam-to-target distance is specified. An axial beam orientation is the default. If another orientation is to be used, then two rotation angles are also specified. For example, in Fig. 7 the beam is first rotated in the polar direction -45° (from top of head toward the right ear) and, secondly, rotated azimuthally -60° (toward the front of the head). In addition to neutron and gamma spectral and angular characteristics, the "expert" source file can contain beam delimiters, shielding, and other structure that are all rotated into place. The beam is rotated rather than the patient because it would be confusing to transform patient coordinates and then combine edit results.

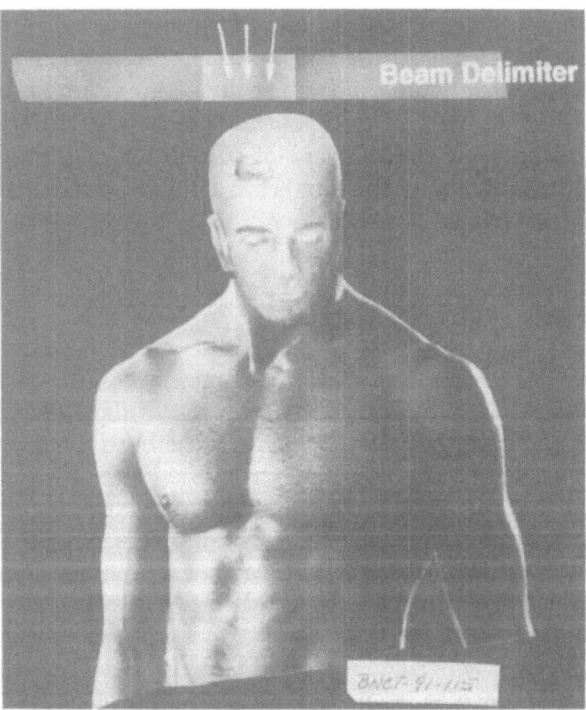

Figure 6. *Axial beam orientation.*

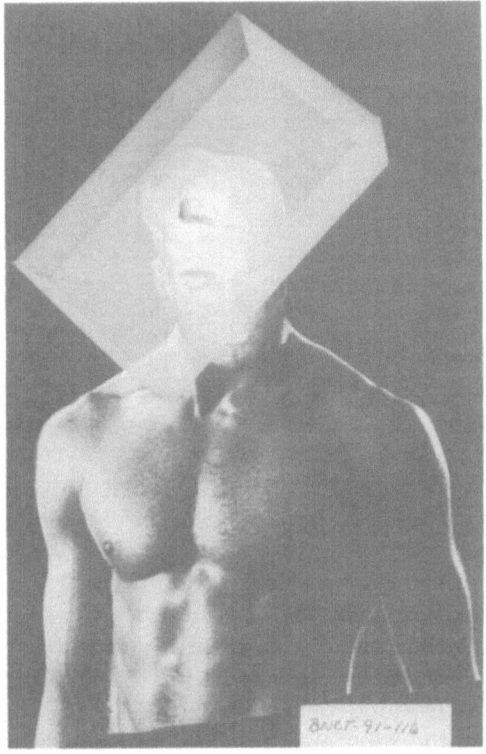

Figure 7. *Oblique beam orientation.*

Thus, Case A represents an unoptimized irradiation plan and Case B re-presents a first step towards an "optimized" plan. It was determined, after performing a few initial beam orientations, that beam angles of (-90°, -60°) had the best possibility. The tumor extent, expressed as depth in tissue along beam centerline, was 4.4-7.5 cm for the axial beam and 3.1-6.2 cm for the oblique beam.

The best metrics useful for determining "figures of merit" for BNCT are presently unknown. The microdosimetry aspects of BNCT are important. Dose and RBE are terms having limited use, and experience from conven-tional planning will be only slightly helpful. "Therapeutic Gain" (TG) (defined in Table 2) will be used as a metric for this trial. The RBE values are not necessarily those applicable to humans, but were determined from cell and animal irradiations at BMRR. The $^{14}N(n,p)^{14}C$ RBE was developed by Archambeau[10] for pig skin, the $^{10}B(n,\alpha)^{7}Li$ RBE was developed by Gabel et al.[11], for Chinese hamster (V79) cells, and the RBE value for the "fast" neutron component of the BMRR epithermal neutron beam was developed by Gavin[12] from canine studies. The RBE value of 0.5 for $^{10}B(n,\alpha)^{7}Li$ in healthy tissue is an upper limit also obtained from Gavin's studies. Of course, this is

Table 2. *Therapeutic Gain (definition)*

T.G. = $\dfrac{\text{RBE dose in tumor}}{\text{Maximum RBE dose in healthy brain}}$

Assuming:

Component	RBE
$^{10}B(n,\alpha)^7Li$	
- tumor	2.3
- healthy brain	0.5
$^{14}N(n,p)^{14}C$	2.7
Fast neutron	5.0
Gamma	1.0
Other	1.0

not a true RBE, but is a factor that accounts for microdistribution of the boron compound, as well as differences in biological effect compared to gamma radiations.

The volume edit capability of the computational system is used to get an integration over the entire target-tissue volume. It is literally vital to include this entire volume since an ignored small volume can result in complete failure of the therapy. Fig. 8 shows this volume edit for the four cases considered. Fig. 8 is similar to a dose/volume histogram except TG and a slightly different format is used to show the relationship of the four cases. The line plot for each case is ordered vs. volume percent from minimum to maximum. Thus, the entire target tissue volume experiences a TG of at least 0.6 for the axial isotropic beam. The plan, for this case, based on TG, is poor. About 60%

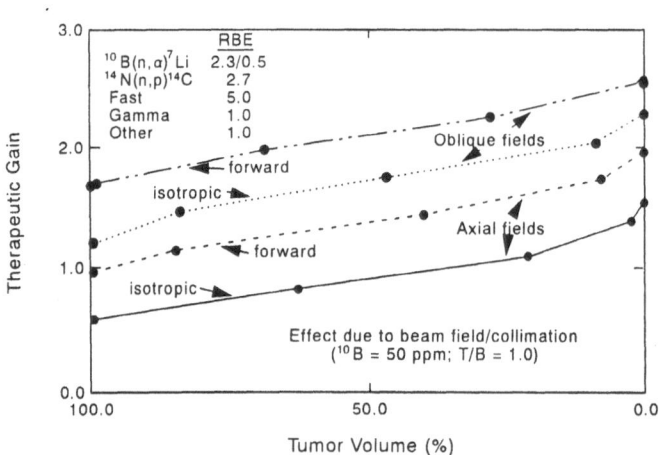

Figure 8. *Therapeutic gain calculated for four treatment cases.*

of the volume has a TG less than 1.0 and only approximately 10% of the volume has a significant TG (greater than 1.2). The axial forward beam is significantly better (based on TG) than the axial isotropic beam, but still only approximately 70% of target tissue volume sees a significant TG.

As common sense would dictate, the oblique beam application provides a very significant increase in TG. Again, the forward-directed beam is clearly superior to the isotropic beam. A forward (monodirectional) beam will generate twice the peak thermal neutron flux in tissues compared to an isotropic beam having identical incident flux intensity. This is merely because the neutron density is double for the forward beam. The relative "fast" neutron component will then be one half that of the isotropic beam for the identical incident neutron energy spectrum for the same delivered boron dose. Because this fast component is so damaging to healthy tissue (RBE ≈ 5.0), TG is dramatically affected. The surface shape of tissue under the beam also affects TG and the forward beam is at an advantage when this shape is not flat. The flux incident on the more distant surface for the forward beam is the same as for the near surface, but falls off for the isotropic beam because of divergence.

This optimization study ceased after these four cases because this was just an exercise. Much more could be done toward optimization with an actual plan. For example, beam aperture size could be varied and multiple fields could be considered. A significantly better plan for this trial case could surely be obtained. Planning will not be as easy as it was for the exercise since one cannot target only the tumor and ignore the occult spread of malignant tissue to distant sites. The target volume to be used is set by the radiation oncologist and could be tumor, tumor plus a certain distance, or even the entire brain. Whatever the strategy, it is clear that a powerful set of planning tools used in the optimization studies will have a dramatic affect on the outcome of initial BNCT trials and later applications.

ACKNOWLEDGMENT

This study was performed under the auspices of the U. S. Department of Energy, Office of Energy Research, under DOE Field Office, Idaho, Contract No. DE-AC07-76ID01570.*

---

* The submitted publication has been authored by a subcontractor of the U.S. Government under DOE Contract No. DE-AC07-76ID01570. Accordingly, the U.S. Government retains a nonexclusive, royalty-free license to publish or reproduce the published form of this contribution, or allow others to do so, for U.S. Government purposes.

REFERENCES

1. Wheeler, F.J.; Griebenow, M.L.; Wessol, D.E.; Nigg, D.W. Analytical modeling for neutron capture therapy. *Strahlenther. Onkol.* 165, 186-188 (1989).

2. Wheeler, F.J.; Nigg, D.W. Three-dimensional BNCT dosimetry analysis in-phantom for the BMRR and PBF epithermal-neutron beams. *Nuclear Science and Engineering Journal*; accepted for publication (March 1991).

3. Cohen, E.; Lyche, T.; Riesenfeld, R. Discrete B-splines and subdivision techniques in computer-aided geometric dDsign and computer graphics. *Computer Graphics and Image Processing* 14, 87-111 (1980).

4. Utah Alpha_1 Project. *Alpha_1 Users' Manual.* Department of Computer Science, University of Utah, Salt Lake City, UT (1986).

5. Peterson, J.W. PRT - A high quality image synthesis system for B-spline surfaces. *Master's Thesis*; University of Utah (1988).

6. Wessol, D.E.; Wheeler, F.J. Methods for creating and using free form geometries in Monte Carlo particle transport; to be published.

7. Kalvin, A.D. Segmentation and surface-based modeling of objects in three-dimensional biomedical images. *Ph.D. Dissertation*; New York University, March 1991.

8. Wheeler, F.J. Monte Carlo techniques for neutron capture therapy. *Int. Workshop on Macro and Microdosimetry and Treatment Planning for Neutron Capture Therapy.* Plenum Press, New York, in the press (1992).

9. Wheeler, F.J.; Parsons, D.K.; Rushton, B.L.; Nigg, D.W. Epithermal-neutron beam design for neutron capture therapy at the PBF and BMRR reactor facilities. *Nucl. Technol.* 92, 106-117 (1990).

10. Archambeau, J.O. Swine Skin: A model to evaluate dose recovery from difference radiations. In: *Clinical Aspects of Neutron Capture Therapy.* Fairchild, R.G.; Bond, V.P.; Woodhead, A.D., eds., Plenum Press, New York, 9-20 (1989).

11. Gabel, D.; Foster, S.; Fairchild, R.G. The Monte-Carlo simulation of the biological effect of the $^{10}B(n,\alpha)^7Li$ reaction in cells and tissues and its implication for boron neutron capture therapy. *Radiat. Res.* 111, 14-25 (1987).

12. Gavin, P.R.; Wheeler, F.J.; Huiskamp, R.; Siefert, A.; Kraft, S.; DeHaan, C. Large animal model studies of normal tissue tolerance using an epithermal neutron beam and borocaptate sodium. In: *Boron Neutron Capture Therapy: Toward Clinical Trials of Glioma Treatment.* Gabel, D.; Moss, R.L., eds., Plenum Press, New York, 197 - 209 (1992).

BOROCAPTATE SODIUM (BSH) PHARMACOKINETICS IN

GLIOMA PATIENTS

Heinz Fankhauser[a], Giuseppe Stragliotto[a],
and Pascal Zbinden[b]

[a]Department of Neurosurgery, CHUV
[b]Institute of Analytical Chemistry, University
CH-1005 Lausanne, Switzerland

INTRODUCTION

In a previous communication we have reported the results of a biodistribution study of intravenous borocaptate sodium (BSH) administered to 28 patients undergoing craniotomy for various intracranial tumors[10]. The measured boron concentrations in tumor, blood, and normal cranial tissues led to the tentative conclusion 1) that BSH readily enters most malignant intracerebral tumors, 2) that some retention of the drug occurs in these tumor, leading to an improvement of the tumor-to-blood ratio over the first 18 hours, and possibly later, and 3) that BSH is virtually excluded from normal brain.

Our ongoing clinical investigation now extends these preliminary findings to 53 patients with intracranial tumors, 26 of whom were found to harbor high and low grade gliomas of various histological subtypes. These 26 patients form the basis for the present report of gross biodistribution of BSH.

PATIENTS AND METHODS

Patients undergoing craniotomy for a likely intracranial neoplasm on computed tomography (CT) or magnetic resonance imaging (MRI) entered the study after informed consent was obtained. An isotonic solution of 95% ^{10}B enriched BSH/kg (initially Callery Chemical, Pittsburgh, USA and later Centronic, New Addington, UK), diluted to a total of 50 ml was administered intravenously over 1 hour. During craniotomy 2 to 72 hours post administration, tumor tissue, blood, and urine were collected in all cases. Additional nor-

mal tissues, including skin, periosteum, muscle, bone, dura, peritumoral normal brain, and CSF were obtained whenever possible. Solid tissues were immediately dip frozen in isopentane and stored at -80°C. Blood was centrifuged
and the plasma, urine, CSF and cyst fluids were stored at -80°C. Boron assessment was performed with Inductively Coupled Plasma - Atomic Emission
Spectroscopy (ICP-AES). Some small samples unsuitable for this method were
analyzed with Quantitative Neutron Capture Therapy (QNCR). These
methods have been described earlier. All final analyses of liquid samples
were done by ICP-AES. Whole blood boron concentrations were calculated
from plasma concentrations by assuming a constant hematocrit of 40%. All
analyses by ICP-AES were repeated 3 times on the same specimen, and the
mean value was used. The results were accepted only when the 3 values were
within 5% of the lowest value. With QNCR, all readings of track densities in
each specimen and standards were done on 10 medium power microscopic
fields either by manual counting or automatic densitometry image analysis.
Means were used as final result. For most tumors and some normal tissues,
several fragments were analyzed. Since necrotic tumor tissue and normal
brain consistently showed lower boron concentrations than highly viable
tumor regions, the values from the fragments with the highest boron concentration are reported in Table 1.

RESULTS

Out of 53 patients with intracranial neoplasms, 26 gliomas were found at
craniotomy, namely 15 glioblastomas, 4 anaplastic astrocytomas, 6 astrocyt-

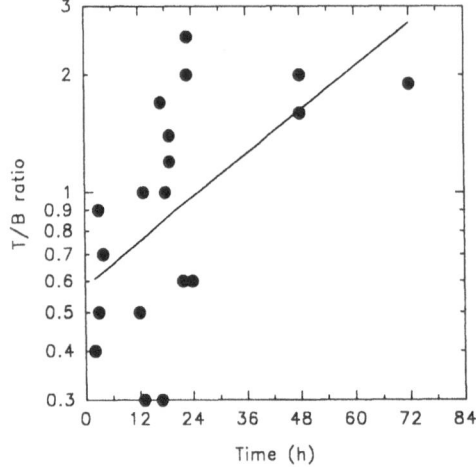

Figure 1. *Semi-logarithmic plot of tumor-to-blood ratio versus interval from
BSH administration in 19 patients with a high grade glioma. The correlation
coefficient of the regression curve is 0.5667.*

Table 1

Boron concentrations (µg/g) in tumor and normal tissues and fluids from 26 patients

Case	Histology	Dose mg^{10}B /kg	Interval (hr)	Tumor (ppm)	Blood (ppm)	Ratio[#] T/B	Brain (ppm)
34/ 1*	Glioblastoma	5	2	6.4	17	0.4	-
43/ 2	"	5	3	7.2	16	0.5	-
52/ 3	"	14	4	8.8	12.7	0.7	1.5
29/ 4*	"	5	12	4	8.7	0.5	-
31/ 5	"	5	13	3.1	3	1	-
36/ 6	"	5	13	1.7	5.2	0.3	0.5
1/ 7	"	2.5	17	1	0.6	1.7	0.2
8/ 8	"	5	17	1	3.1	0.3	-
48/ 9*	"	5	22	1.7	2.9	0.6	-
10/10*	"	5	23	2.7	1.3	2	1
38/11*	"	5	23	7.5	3	2.5	0.5
57/12	"	5	24	5.5	9	0.6	-
55/13	"	15	48	5	3.1	1.6	-
xx/14	"	25	48	7.1	3.4	2	-
56/15	"	18	72	3.4	1.8	1.9	-
28/16	Astrocytoma grade III	5	3	10	11	0.9	-
18/17	"	5	18	2.6	2.6	1	-
11/18	"	5	19	1.5	1.3	1.2	1
16/19*	"	5	19	2.2	1.6	1.4	0.5
41/45/20	Astrocytoma grade II	5	4	5.4	15	0.4	-
39/21	"	5	12	5.5	9	0.6	0.3
15/22	"	5	16	1.5	1.5	1	-
49/23	"	14	18	2.5	4.8	0.5	-
24/24	"	5	19	0.3	0.5	0.6	0.2
14/25	"	5	25	0.4	0.6	0.7	0.4
13/26	Oligodendroglioma	5	27	0.2	0.9	0.2	0.2

*recurrent tumors, #tumor-to-blood ratio

omas grade II, and 1 oligodendroglioma grade II. Six of these patients were operated for recurrent tumors. Histological diagnosis, amount of ^{10}B administered intravenously, and post BSH administration interval to tissue harvesting during craniotomy are detailed in Table 1. Tumor, blood, and normal brain values for ^{10}B, as well as tumor-to-blood ratios are listed in the same table.

The concentration of boron in blood at the time of tumor sampling varied according to the delay from injection. The values measured at 24 hours were roughly 4 times lower than at 3 hours in patients with an identical BSH dose. Patients receiving a larger dose of boron had a higher final boron amount in blood. Blood boron concentrations were not related to grade of malignancy.

The maximal tumor boron concentrations were observed in high grade intracerebral tumors. The highest value, 10 ppm, was measured 3 hours after administration; after 18 hours, the maximum concentration was 2.6 ppm, after 24 hours 7.5 ppm, and after 48 h hours 7.1 ppm. A tumor-to-blood ratio above 1 was obtained in 8 out of 19 high grade intracerebral tumors; in all but one this was seen 17 hours or later after injection (Fig. 1).

Normal brain values reached a maximum of 1.5 ppm. In the 7 cases with high grade gliomas where normal brain was available, the tumor-to-brain ratio was 1.5 in one, and above 2.5 in the remainder. Boron was measured in other normal tissues whenever available during craniotomy. Concentrations were high in skin and temporal muscle and reached roughly the values found at the same time in high grade gliomas. Boron was low in bone and very low or undetectable in CSF.

In low grade intracerebral tumors, the highest tumor boron concentration of 5.5 ppm was measured after 12 hours delay. The 3 values at 19 hours and later were below 0.5 ppm. No tumor-to-blood ratio above 1 was found. Normal brain values measured in 4 patients with low grade intracerebral tumors were 0.4 ppm or below.

DISCUSSION

At the present time BNCT is primarily considered for malignant gliomas[7,8]. The reasons may be mainly historical, but it can be justified by the fact that there has been no decisive progress in the treatment of this devastating disease, which has an extremely low tendency for distant metastases despite invariable local recurrence. Low grade gliomas, regardless of their slower growth rate, basically share the same characteristics and show a propensity for transformation into higher grade gliomas.

The amount of BSH administered to our patients is up to 10 times less than in the ongoing treatment study in Japan[7,8]. This option could be chosen since current technology allows to measure boron concentrations in virtually all tissues even with a low dose. Therefore, the non-therapeutical nature of our trial and the high price and irregular supply dictated initial small BSH administration.

Determination of small amounts of boron in biological fluids and tissues is difficult, and potentially reliable methods have been established only recently[1,2,3,6,11,12,13]. The accuracy of QNCR has been found to be around $\pm 30\%$, with a tendency towards overestimation of boron concentrations. In our hands, ICP-AES is simpler and clearly more reliable than QNCR, with an accuracy of approximately $\pm 5\%$. Our initial specimen were analyzed exclusively with QNCR, but most of these, including all tumor specimen, were analyzed again with ICP-AES. Later specimen were all analyzed by ICP-AES or both methods. We consider QNCR as useful in situations where histological correlation is more important than knowledge of absolute concentrations.

BSH enters gliomas at an early time, since maximum tumor concentrations were measured 2-3 hours after infusion. Initial tumor concentrations nevertheless stay below blood concentrations. After 15-18 hours this relation can be reversed. In high grade gliomas boron clearance from tumor is slightly slower than from blood, leading to an improved tumor-to-blood ratio. There is some indication that the tumor-to-blood ratio further improves at 24 and 48 hours (Fig. 1). This question will be clarified by including more patients at these intervals. At the doses we injected, the best tumor-to-blood ratio did not exceed 2.5. With an injection of 5 mg B/kg, we found a maximum of 7.5 ppm B in malignant tumors at 24 hours. By increasing the injected amount to 25 mg B/kg, we observed a similar concentration even at 48 hours. Present data nevertheless do not allow to determine the precise relation between injected dose and tumor concentration.

Haritz et al.[5,6] have reported on BSH distribution in 8 patients with malignant gliomas in a study similar to ours. Their findings confirm significant accumulation of boron in malignant tumor tissue. Using QNCR, they were able to provide correlation between histology of various tumor regions and local boron concentration, and to demonstrate a pronounced heterogeneity. Low concentrations appeared to correlate with necrotic, microcystic, and relatively tumor free areas. As in our patients, the tumor-to-blood ratio improved over 24 hours.

From our study of macroscopic boron distribution, low grade intracerebral tumors accumulate low amounts of the isotope and do not achieve a positive tumor-to-blood ratio. This would make them unsuitable for BNCT, unless further analyses provide evidence for a selective boron accumulation at the cellular or subcellular level.

In 11 patients we obtained "normal" brain matter for analysis. Although this tissue was macroscopically and microscopically tumor free, it usually stemmed from a region adjacent to the tumor. Therefore, this represented edematous brain with grossly intact BBB. Despite this, very low boron concentrations were found in these specimens, not exceeding 1.5 ppm. This was observed for time intervals between 4 and 27 hours. The boron concentration in CSF was invariably low, compared to tumor or blood values. By analogy with other studies on delivery of drugs from blood to CSF, we concluded that no selective transport occurs across the ependymal lining.

A drawback of our tumor sampling method is that in each case we were able to measure only a single or a few close time-points, which cannot reflect the dynamic event of compound accumulation and elimination over time. Since there is a heterogeneity of accumulation not only from patient to patient, but also from one tumor area to another, the absolute values and the ratios between tumor and normal tissues in our cases are to be considered as maxima under optimal conditions. Histological comparison showed that the highest B concentrations for tumors were found in highly neoplastic zones, i.e. in well vascularized, anaplastic tissue, whereas lower boron values were found in necrotic areas. In terms of therapeutic trials, necrotic areas and the macroscopically neoplastic bulk are likely to be removed at operation preceding BNCT. For this reason, and whenever several tumor specimens were available from the same patient, we reported the highest boron concentration measured in that tumor.

Further investigations in high grade gliomas have to clarify whether there is any significant improvement of the tumor-to-blood ratio later than 24 hours after injection. If this is the case, careful evaluation of relative merits of high absolute tumor boron concentration, which will be lost at longer intervals, versus high tumor-to-blood ratio has to be carried out before any definite decision is taken concerning the administration schedule. As evident from our results, at shorter injection intervals BSH remains confined to the tumor region, but it is not known whether the drug will move into brain edema and follow the bulk flow of brain water towards the ventricles if sufficient time is allowed. Determination of slow diffusion of BSH into tumor-free brain is a mandatory step in our ongoing investigations, since patients are likely to be treated by fractionated BNCT and hence will receive repeat BSH infusions at 24 or 48 hours intervals.

In low grade intracerebral tumors there is less histological heterogeneity than in malignant forms, and the variability of the B concentration was found to be narrower, but this may be of limited significance since absolute tumor concentrations were lower and tumor-to-blood boron ratios were usually below 1. Unless methods become available which allow precise measurements of B concentrations at the cellular and subcellular level, no definite prediction can be made about the chances of low grade intracerebral tumor to benefit from BNCT.

Comparison of our results with available publications is limited. Hatanaka used BSH in doses 8-16 times higher than ours, and obtained a mean maximum tumor-to-blood ratio of 1.1. However, no detailed information about the performance of chemical analysis are given[7,8]. From our quality control experiments we conclude that in our setting the analytical methods give raise to variations that usually do not exceed 10%, which can be considered as acceptable[11]. Nevertheless, with tumor-to-blood ratios around 1, this variability may turn a positive ratio into a negative one and vice versa. Therefore, only values above 1.5 should be considered as a definitely positive tumor-to-blood ratio. Hatanaka's protocol differed from ours since he examined tumors during re-craniotomy, shortly after a previous debulking procedure, at which time the tumor bed may have contained only few growing cells and therefore less boron. Despite this, Hatanaka observed tumor-to-blood ratios above 3 in certain instances, a figure which was never reached in our intact malignancies[8].

The aim of this study was to obtain data on biodistribution of BSH, to be used for the design of a preliminary BNCT trial with an epithermal neutron beam. Assuming that a positive tumor-to-blood ratio and a uniform tumor concentration of 30 ppm B are needed to benefit from a significant therapeutic gain from the boron neutron capture reaction over the nonspecific radiation dose components, our findings would indicate that a dose of 80 to 100 µg BSH/kg should be given one day before irradiation. Recent dose calculations taking into account the particular distribution pattern of BSH and the ensuing vascular sparing effect indicate that with the above tumor concentration a tumor-to-blood ratio of only 0.5 would be sufficient to deliver a physical radiation dose to the tumor which exceeds the dose to the capillary endothelium[4]. As far as BSH is concerned, high absolute tumor boron concentration therefore appears to contribute more to the desired dose distribution than a high tumor-to-blood ratio. An optimal treatment schedule therefore may require radiation as early as 3 hours after BSH administration. Preliminary clinical evidence for the validity of this assumption has recently been observed in dog irradiation studies[4].

ACKNOWLEDGEMENTS

The authors would like to thank Prof. Patrick R. Gavin, Washington State University, Pullman, WA for his critical review of the manuscript. This work has been supported by the Fonds National de la Recherche Scientifique, the Ligue Suisse contre le Cancer, the Ligue Vaudoise contre le Cancer, and the Commission of the European Communities.

REFERENCES

1.   Barth, R.F.; Adams, D.M.; Soloway, A.H.; Mechetner, E.B.; Alam, F.;
      Anisuzzaman, A.K.M. Determination of boron in tissues and cells
      using direct-current plasma atomic emission spectroscopy. *Anal.
      Chem.* 63, 890-893 (1991).
2.   Bauer, W.F.; Johnson, D.A.; Steele, S.M.; Messick, K.; Miller, D.L.;
      Propp, W.A. Gross boron determination in biological samples by In-
      ductively Coupled Plasma Atomic Emission Spectroscopy. *Strahlen-
      ther. Onkol.* 165, 176-179 (1989).
3.   Gabel, D.; Holstein, H.; Larsson, B.; Ericson, G.; Sacker, D.; Som, P.;
      Fairchild, R.G. Quantitative neutron capture radiography for study-
      ing the biodistribution of tumor-seeking boron-containing compounds.
      Cancer Res. 47, 5451-5454 (1987).

4.   Gavin, P.; Wheeler, F.J.; Huiskamp, R.; Siefert, A.; Kraft, S.; DeHaan, C. Large animal model studies of normal tissue tolerance using an epithermal neutron beam and borocaptate sodium. In: *Boron Neutron Capture Therapy: Toward Clinical Trials of Glioma Treatment.* Gabel, D.; Moss, R.L., eds., Plenum Press, New York, 197 - 209 (1992).

5.   Haritz, D.; Piscol, K.; Gabel, D. Distribution of BSH in patients with malignant glioma. In: *Progress in Neutron Capture Therapy for Cancer.* Allen, B.J.; Moore, D.E.; Harrington, B.V., eds., Plenum Press, New York, 557-560 (1992).

6.   Haritz, D.; Gabel, D.; Klein, H.; Huiskamp, R.; Pettersson, O. BSH in patients with malignant glioma. Distribution in tissues, comparison between BSH concentration and histology. In: *Boron Neutron Capture Therapy: Toward Clinical Trials of Glioma Treatment.* Gabel, D.; Moss, R.L., eds., Plenum Press, New York, 163 - 174 (1992).

7.   Hatanaka, H. Experience of boron-neutron capture therapy for malignant brain tumors with special reference to the problems of postoperative CT follow-up. *Acta Neurochir. Suppl.* 42, 187-192 (1988).

8.   Hatanaka, H.; Amano, K.; Kanemitsu, H.; Ikeuchi, I.; Yoshizaki, T. Boron uptake by human brain tumors and quality control of boron compounds. In: *Boron-Neutron Capture Therapy for Tumors.* Hatanaka, H., ed., Nishimura, Niigata, Japan (1986).

9.   Kraft, S.L.; Gavin, P.R.; DeHaan, C.E.; Bauer, W.F.; Ary, T.E. The biodistribution of boron in normal canine tissues following borocaptate sodium administration and the effect of plasma exchange. In: *Progress in Neutron Capture Therapy for Cancer.* Allen, B.J.; Moore, D.E.; Harrington, B.V., eds., Plenum Press, New York, 489-492 (1992).

10. Stragliotto, G.; Fankhauser, H. Biodistribtution of boron sulfhydryl (BSH) in patients with intracranial tumors. In: *Progress in Neutron Capture Therapy for Cancer.* Allen, B.J.; Moore, D.E.; Harrington, B.V., eds., Plenum Press, New York, 551-556 (1992).

11. Stragliotto, G.; Zbinden, P.; Pettersson, O.; Fankhauser, H. Biodistribution of boron sulfhydryl (BSH) in humans: A quality control of analytical methods. In: *Progress in Neutron Capture Therapy for Cancer*; Allen, B.J.; Moore, D.E.; Harrington, B.V., eds., Plenum Press, New York, 545-548 (1992).

12. Tamat, S.R.; Moore, D.E.; Allen, B.J. Determination of the concentration of complex boronated compounds in biological tissues by Inductively Coupled Plasma Atomic Emission Spectroscopy. *Pigm. Cell Res.* 2, 281-285 (1989).

12. Yoshino, K.; Kajiyama, Y.; Honda, T.; Mori, Y.; Honda, C.; Ichihashi, M.; Mishima, Y. A trial to improve the analysis of boron in biological materials. *Pigm. Cell Res.* 2, 286-290 (1989).

BSH IN PATIENTS WITH MALIGNANT GLIOMA:

DISTRIBUTION IN TISSUES, COMPARISON BETWEEN

BSH CONCENTRATION AND HISTOLOGY

Dietrich Haritz[a], Detlef Gabel[b], Harold Klein[c],
René Huiskamp[d], and Orn-Anong Pettersson[e]

[a]Dept. of Neurosurgery
(Director: Prof. Dr.med. K. Piscol)
ZKH St. Jürgenstraße
D(W)-2800 Bremen, F.R. Germany

[b]Dept. of Chemistry
University of Bremen
D(W)-2800 Bremen, F.R. Germany

[c]Institute of Clinical Neuropathology
ZKH Ost
D(W)-2800 Bremen, F.R. Germany

[d]Netherlands Energy Research Foundation (ECN)
NL-1755 ZG Petten, The Netherlands

[e]Institute of Radiation Science
University of Uppsala
S-751 21 Uppsala, Sweden

INTRODUCTION

The treatment of malignant brain tumors, especially glioblastoma multiforme, presents great problems to the therapeutic disciplines. Despite technical advances in the surgical management and the perioperative treatment, a
successful surgical approach is improbable, due to the diffuse and infiltrating
growth of the tumor. From the standpoint of the radiotherapist the low
tolerance of the normal brain tissue limits the delivery of tumor effective

Boron Neutron Capture Therapy, Edited by D. Gable
and R. Moss, Plenum Press, New York, 1992

radiation dosages. Finally, chemotherapy has shown to be ineffective in bringing about lasting remission of malignant glioma (WHO III/IV) or even in control tumor growth. Even the current aggressive multimodality therapy did not result in an average survival time of longer than 12 months. One is still faced with an infaust and very depressing prognosis for the patient suffering from a glioblastoma multiforme.

Based on its different theoretical aspects, there are many convincing facts that suggest that Boron Neutron Capture Therapy (BNCT) could be developed to an effective treatment modality for malignant brain tumors[1]. The group in Bremen started a clinical phase-I-study to investigate the pharmacokinetics, biodistribution and possible toxicity of the boron compound $Na_2B_{12}H_{11}SH$ (BSH). The study is part of the European Collaboration on BNCT and carried out in cooperation with the research centers in Petten and Uppsala.

Table 1

Blood parameters tested

Bilirubin	Protein
SGPT (glutamate pyruvate transaminase)	Hemoglobin
	Erythrocytes
SGOT (glutamate oxalacetate transaminase)	Leukocytes
	Platelets
GGT (gamma-glutamyl transpeptidase)	Clotting system
Cholinesterase	Sodium
AP (alkaline phosphatase)	Potassium
Amylase, Lipase	Chloride
LDH (lactate dehydrogenase)	Calcium

PROTOCOL

A coordinated protocol has been established within the European Collaboration. BSH is infused during the course of 1 hour and blood and urine samples are taken at predetermined intervals. During the operation tissue samples are obtained. These include: skin, muscle, bone, dura mater, cerebrospinal fluid (CSF), normal brain (if available) and tumor tissue. These different types of tissue will eventually be exposed to the neutron beam. Consequently, informations about the concentration of the boron compound is of utmost importance for the prediction of healthy tissue damage and tumor control.

The patient eligible for the study, which is authorized by the local Ethical Board, must have a presumed high grade glioma by CT, MRI or angiography. The patients are well informed about their presumed malignant tumor and

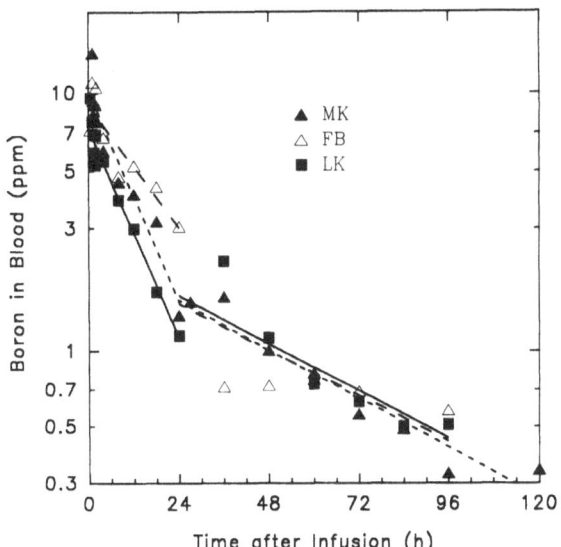

Figure 1. *Pharmacokinetics of BSH. Data for 3 of the 8 patients are shown. Observation starts at the initiation of infusion. The different symbols represent patient MK (open circles), patient LK (closed triangle), and patient EA (open triangle). The lines represent the regression curves for each patient for times of 1-24 hours, and for ≥ 24 hours respectively.*

the non-therapeutic aim of the study and participate on a voluntary basis. Patients with reduced function of liver, lung, kidneys and cardio-vascular system are excluded, as are those with major endocrinological disturbances. Table 1 shows the list of the clinical chemical parameter, which are measured prior to administration and again on the first, second, seventh and fourteenth day after the administration, in order to evaluate possible toxic side effects of BSH. The control group includes patients undergoing the same surgical treatment, but because of a non-malignant brain tumor. These patients fulfill the same preconditions as those entering the study.

The boron compound, dissolved in 50 ml sterile saline, is infused through a central venous catheter. The infusion is started 3, 6, 12, 18 or 24 hours before the operation. For the patients in this paper, a maximum amount of 15 mg BSH per kg body weight was permitted. Blood is taken at predetermined intervals up to 120 hours after the start of the infusion. Removed tissues and fluids are frozen rapidly in liquid nitrogen and stored in a freezer at -23°C. Urine samples are taken after 1, 2, 4, 8, 12, 18, 24 hours, and then with intervals of 12 hours up to 96 hours. Aliquots of each amount of urine are stored in a freezer at -23°C.

BORON ANALYSIS AND HISTOLOGICAL INVESTIGATIONS

The concentration of boron in blood and tissues is measured by quantitative neutron capture radiography (QNCR)[2], that in urine and CSF by inductively coupled plasma-atomic emission spectroscopy (ICP-AES) at ECN Petten. Because of a necessary cross calibration between the different boron determination methods, blood samples of some patients are measured by ICP-AES as well. For QNCR, the samples are embedded in 3% carboxymethyl cellulose. Cryosections of 50 μm thickness are cut every 1 mm. Standards are prepared from chicken liver, calf brain or human blood by adding known amounts of boric acid. The freeze-dried cryosections are mounted onto Kodak Pathé LR 115 Type 1 track detectors and exposed to around $5 \cdot 10^{12}$ neutrons cm^{-2} at the Neutron Radiography facility in Studsvik/Sweden. The detectors are etched at 60°C in 10% NaOH for around 50 minutes. The etched detectors are evaluated in an image analyzer, consisting of a microscope, a TV camera, a personal computer with appropriate software, and a monitor. With a bone biopsy needle of 2 mm diameter, small tissue samples are obtained from the surface of the frozen material during microtomy, in order to correlate the concentration of boron with the type of tissue. After fixation in 10% formalin, embedding in paraffin, cutting and staining in hematoxyline-eosine, these samples are examined neuropathologically. The histological diagnosis is verified and the type of tissue and density of cells in the sample are described.

RESULTS OF THE TOXICITY INVESTIGATION

None of the 8 patients that entered the study to date, showed any measurable acute toxic side effect of BSH on cardio-vascular, liver, kidney and endocrinological functions or any allergenic reaction. The patients did not complain of any subjective disorders such as nausea or vomiting. The different blood tests compiled in Table 1 were evaluated statistically by using the Wilcoxon test with a 95% level of significance. For the transaminase SGOT, a significant difference was found within the control group for pre- and postoperative values, but not for the BSH group. All other parameters did not differ between the study group and the control group. With 15 mg BSH per kg body weight we found no indication of a pharmacodynamic effect of the boron compound BSH in man.

BORON PHARMACOKINETICS

The boron concentration in blood decreased non-exponentially. When the data were divided arbitrarily in the time intervals 1 to 24 hours, and 24 hours or longer, we found a biphasic kinetic with an apparent first half-life of around 10 hours and an apparent second half-life of around 25 hours. Repre-

Table 2

Summary of the Patients Evaluated in this Study

Code	Age/sex	Weight (kg)	mg B /kg	$t_{1/2}$ 1-24h	$t_{1/2}$ ≥24h	% B excreted in urine
MK	63/f	68	8.4	8.9	44.1	55
LK	71/f	56	10.2	8.4	41.9	66
VO	62/m	83	6.9	8.0	43.0	88
FB	70/m	80	7.1	12.1	43.5	43
EA	69/f	53	10.8	10.5	35.0	94
V	50/m	74	7.7	8.3	39.0	49[*]
RUP	55/f	75	7.6	4.4	11.9	38[**]
VORM	61/m	81	7.1	7.4	26.0	96
average				8.5	35.6	66
± SD				±2.2	±11.3	±23

[*] within the first 48 hours after administration
[**] within the first 36 hours after administration
Half lives are derived from regression analysis of all samples collected between 1 and 24 hours, or collected at 24 hours and longer.

sentative for all patients, Fig. 1 shows the graph of the boron concentration in blood from three patients over the observation time. The initial distribution volume of BSH (Table 2) is smaller than the total body water (which might be approximated to around 65% of the total body weight). Boron is found in the urine shortly after the administration. Further investigations have to clarify the distribution of the compound into deeper compartments with delayed excretion, which is may be controlled by an entero-hepatic pathway.

Around 570 mg ^{10}B has been administered to each of the patients. Over the observation time, only about two thirds of the compound is excreted through the urinary system. Fig. 2 shows that the excretion reaches a limit. Recovery of boron ranges from 38% (patient RUP) to 96% (patient VORM) (Table 2). This strongly indicates that the compound is not be excreted exclusively through the urinary system.

BORON DISTRIBUTION IN TISSUE

The evaluation of the QNCRs consistently showed a highly heterogeneous distribution of BSH in the neuropathologically verified tumor tissue[4]. Parts of the tumor tissue accumulated BSH in very high concentrations. This correlated with the histological finding of dense tumor cells with many pathological vessels. In other tissue samples, adjacent areas less than 1 mm apart, differed by up to one order of magnitude in boron concentration. This corresponded either with spongeous formation of tumor cells with only a few vessels or signs of cellular degeneration. In necrotic areas, BSH uptake was very low. In healthy brain, no significant boron uptake was measured.

Table 3 shows a summary of the eight patients studied so far. The ratio of the boron concentrations in tumor *vs.* blood increased with increasing time between BSH administration and sampling of the tumor tissue. The maximum values of the tumor to blood ratio reached higher levels as well. Many of the tumor samples contained boron in a concentration ratio of tumor to blood greater than 1. For patient FB, only 2 of 19 proven tumor samples dis-

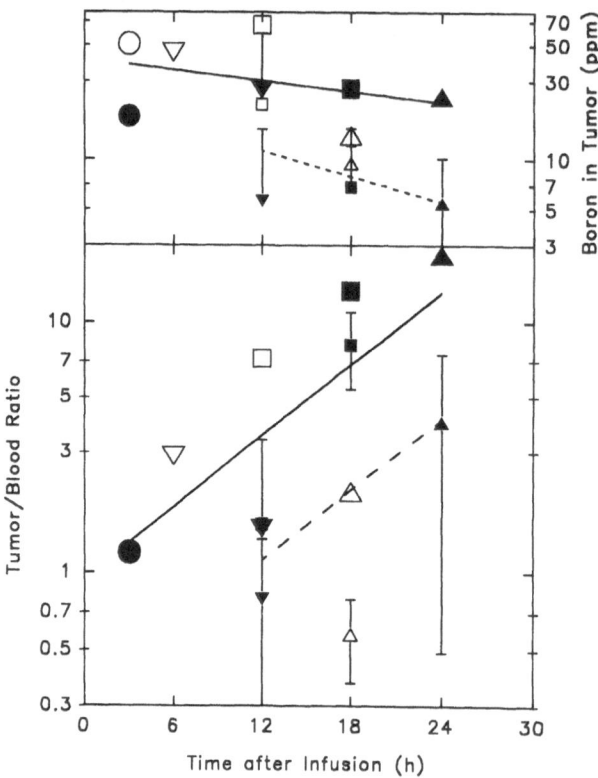

Figure 2. *The graph shows the excretion of boron through the urinary system for 4 of the 8 patients over the observation period of 96 hours. The lines represent exponential functions fitted to the measured points. The patients received 0.9 g BSH (517 mg boron) intravenously. The excreted amounts of boron range between 43% and 96% of the total administered dose.*

played dense accumulation of tumor cells with a corresponding high concentration of BSH. The rest of the tissue samples showed a very spongeous cell formation and many necrotic areas, which might explain why the measured maximum tumor-to-blood ratio was very low.

Fig. 3 summarizes the tumor-to-blood ratio of the boron concentration and the boron concentration measured in tumor tissue as a function of the time between BSH administration and sampling of the tissue. The maximum values are shown for all eight patients. For five of the patients the neuropathological status of the tumor tissue analyzed for boron was verified. For these patients, the average boron values in tumor tissue could be specified. For the other patients, only maximum concentrations could be determined, as no tissue typing was carried out. The initial high uptake of BSH decreased only slowly with time. As the blood was cleared rapidly (see Fig. 1) there was a significant increase with time of the tumor-to-blood ratio. This pertains to both the maximum and the average values. In other tissues eventually exposed to the neutron beam, boron concentration was very low (bone, fat, CSF) or occasionally reached up to blood concentration (skin, muscle) (Table 4).

Table 3

Boron in tumor tissue

Patient	Interval (h)	Concentration ratios of tumor to blood		
		average ± SD	>1	maximum
VORM	3	n.d.	n.d.	1.2
RUP	3	n.d.	n.d.	1.2
V	6	n.d.	n.d.	2.9
EA	12	0.79 ± 0.56	7/14	1.4
VO	12	1.56 ± 1.80	9/20	4.6
MK	18	8.18 ± 2.80	19/19	11.8
FB	18	0.58 ± 0.21	2/19	1.5
LK	24	3.97 ± 3.48	34/35	18.4

Boron concentration ratios of tumor to blood for all patients. The second column represents the time between BSH administration and surgery, the third column the average (with standard deviation) of the concentration ratio of tumor to blood, the fourth column the number of the histologically confirmed tumor biopsies with a concentration ratio greater than 1, and the fifth column the maximum tumor-to-blood boron concentration ratio measured in the proven biopsies (patient EA through LK) resp. surgically removed brain or tumor tissue (patient VORM through V) from the embedded tissue. "n.d.": not determined.

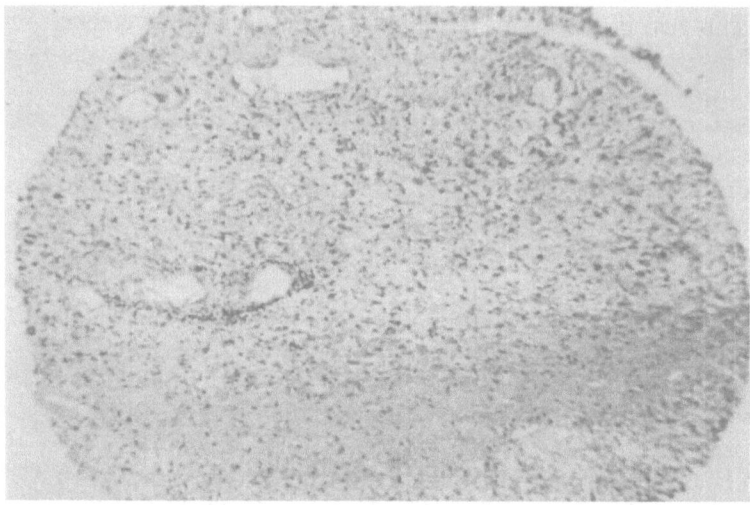

Figure 3. *The graph shows the boron concentration ratio in tumor to blood dependent on the time after the BSH administration (lower part), and the measured boron concentration in tumor tissue (upper part). The lines represent the regression curve of the maximum values of all of the eight patients. The interrupted lines represent the average values of the last five patients, where the obtained biopsies are proven histopathologically. With a longer period between the BSH administration and the tissue sampling an increase of the tumor to blood ratio can be observed, whilst the decrease of the boron concentration in the tumor tissue is very low over the time.*

DISCUSSION

In our study, we show that boron from BSH accumulates in malignant brain tumor tissues. The uptake of the compound is highly heterogeneous, with differences of one order of magnitude and more between adjacent parts separated by distances of a few millimeters and less. In normal brain no significant uptake is measured. A low uptake of the BSH and its dimer BSSB in normal brain tissue after protracted intraperitoneal infusion and favorable melanoma-to-blood and melanoma-to-normal brain concentration ratios in mice has been described by Slatkin[5]. Moreover, they found a retention of the boron compound in the tumor tissue and a decline of the boron concentration in blood over the observation period of 3 days after cessation of the infusion. This would indicate a potential benefit from using BNCT as a high LET radiation treatment for malignant brain tumors in man[5].

Table 4

Boron in normal tissues

Patient	Interval	Concentration ratios of tissue-to-blood		
	(h)	muscle	skin	bone
VORM	3	-	-	0.1
RUP	3	0.5	0.0	0.0
V	6	-	-	0.1
EA	12	0.1	0.0	0.0
VO	12	0.8	0.5	0.1
MK	18	0.5	0.9	0.1
FB	18	0.4	0.3	0.1
LK	24	2.0	1.3	0.3

Boron concentration ratios in tissues (obtained during surgery at predetermined intervals) to blood in all of the eight patients. For patients VORM and V, samples of muscle and skin were not obtained.

The data of our study correspond satisfactorily with the results of maximum boron concentrations in tumor tissue and normal brain, reported by Fankhauser[6]. Also in tumor bearing dogs the uptake of the BSH in normal brain tissue was found to be very low, at least in areas outside the disrupted blood brain barrier, as measured with ICP-AES[7].

When using ICP-AES for boron determination in tissue samples, there exists, however, no possibility to obtain information about the microdistribution of BSH within the evaluated sample. Because of the histological characteristics of a malignant brain tumor, especially a glioblastoma multiforme, a method for boron determination with a good spatial resolution is needed. This can be provided by QNCR. On the other hand, ICP-AES gives some advantages for the evaluation of a large number of samples and seems to be the method with an analytical error smaller than QNCR. The outcome of clinical trials using BNCT as a new treatment modality for glioblastoma would greatly depend on the distribution of boron within the fringe of the tumor, which contains normal tissue and highly malignant tissue in adjacent parts separated without a sharp borderline. The concentration of BSH in those parts of a tumor which cannot be removed satisfactorily by surgery is of utmost importance for the outcome of patients treated by BNCT. This should always be kept in mind because of the unsharp borderline between tumor tissue and normal brain tissue. Therefore we consider knowledge about the

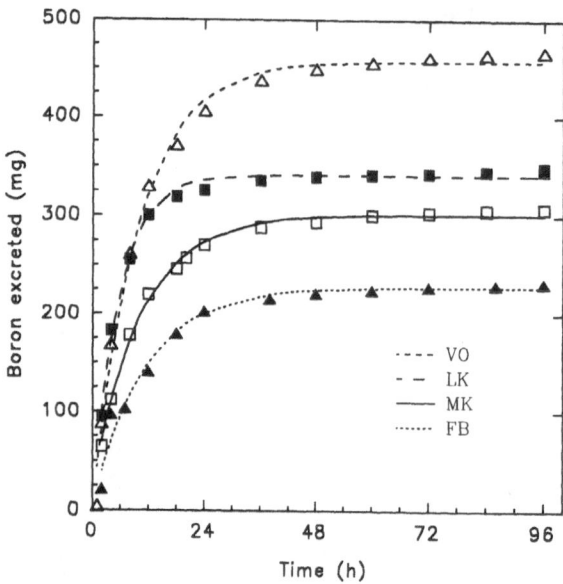

Figure 4. *Hematoxylin-eosine stained tissue biopsy obtained from the cryo-embedded material (the field shown is 1 x 1 mm². A small stripe of normal brain tissue surrounded by tumor tissue with signs of cystic degeneration. On the left side of the specimen a pathological blood vessel.*

correlation between boron distribution and type of tissue as extremely important. With the biopsy needle technique developed by us, in combination with QNCR, there exists a method for obtaining this correlation, with a resolution of down to 0.1 mm. An example of the good histological quality of the tissue thus obtained is shown in Fig. 4.

The investigation about the pharmacokinetics of BSH shows a biphasic kinetic with an excretion mainly through the urinary system. After evaluation of 8 patients, the previous data[3,4] about the course of the boron concentration in blood have been confirmed. It is therefore probable that the compound distributes only into the plasma and interstitial spaces, at least during the first 24 hours. Already shortly after the start of the infusion, boron is found in the urine. Therefore it is unlikely that the first half-life would represent only the distribution phase and the second would correspond to the elimination phase. In the time interval of 2 hours after administration and 120 hours (the end of the observation period), we found two half-lifes, each representing both distribution and elimination of BSH. In contrast to the Lausanne group[6] we found no early half-life of around 1 hour. In the first hour after cessation of the infusion we found, however, a small plateau phase of the boron concentration in blood. Entero-hepatic reabsorption could be an explanation for this finding.

Significant differences of the half-lifes of administered BSH is found between mice, dogs, and humans. In mice, a first half-life of around 1.5 hours and a second half-life of about 12 hours was found[2]. In dogs, a half-life of around 5 hours was found (the reported observation periods are too short to allow the verification of a second half-life)[7,8]. In patients, a much longer first half-life of around 8.5 hours was found, followed by a second one of around 35 hours. The pharmacokinetics of BSH seems to be different from species to species.

The evaluation of the boron distribution in blood and the recovery of the compound in urine confirms that, differing from patient to patient, an amount of up to 62% of the compound is excreted through an unknown, perhaps entero-hepatic pathway, or retained in the body. The fact that not all boron is excreted through the urinary system indicates the existence of an additional excretion pathway, which is unknown so far, but might be represented by the above-mentioned hepatic pathway. Alternatively, prolonged retention in other tissues might be responsible for the observed discrepancy between urinary excretion and amount administered. A significant concentration of BSH in other tissues and organs has been reported for dogs by Kraft[7].

With the amounts of BSH used in this study, we observed no toxic side effect of the compound. It may be assumed that BSH is a substance with no or very small pharmacodynamic effect. Also the Lausanne study group observed no obvious toxic side effect related to the administration of BSH up to dosages of about 50 mg BSH/kg[6].

When taking into account blood clearance, boron uptake in tumor tissue and concentration ratios, the interval of around 18 hours between administration and the planned irradiation might be optimal. In contrast to the results in tumor bearing dogs, we found an increasing tumor to blood concentration ratio of BSH up to an interval between 18 and 24 hours after intravenous administration of the compound. The initial uptake of the compound in tumor tissue is consistently high in both tumor bearing dogs and human malignant brain tumors. However, the decline of the tumor boron concentration in dogs with spontaneous brain tumors is faster than the decline in human malignant brain tumors. This might be the reason why Kraft did not found an increase of the tumor-to-blood concentration ratio over the observation period[7]. It must be borne in mind that none of the specified tumors in dogs are histopathologically comparable with a human glioblastoma multiforme[7]. In contrast to the dog tumor study, we did find an increase of the tumor-to-blood ratio of boron with time. In contrast to the result in tumor-bearing dogs, the amount of boron in the tumors of patients did not decline in parallel with the amount of boron in blood.

ACKNOWLEDGMENTS

Financial support from the Senator for Health of Bremen and the Commission of the European Communities is gratefully acknowledged.

REFERENCES

1. Hatanaka, H., ed. *Boron-Neutron Capture Therapy for Tumors.* Nishimura, Niigata/Japan (1986).
2. Gabel, D.; Holstein, H.; Larsson, B.; Gille, L.; Ericson, G.; Sacker, D.; Som, P.; Fairchild, R.G. Quantitative neutron capture radiography for studying the biodistribution of tumor-seeking boron-containing compounds. *Cancer Res.* 47, 5451-5456 (1987).
3. Haritz, D.; Gabel, D.; Piscol, K. The distribution of BSH in patients with malignant glioma. In: *Progress in Neutron Capture Therapy for Cancer.* Allen, B.J.; Moore, D.E.; Harrington, B.V., eds., Plenum Press, New York, 557-560 (1992).
4. Haritz, D.; Gabel, D.; Klein. H.; Piscol, K. Clinical investigation in boron neutron capture therapy (BNCT). Pharmacokinetics, biodistribution and toxicity of Na2B12H11SH (BSH) in patients with malignant glioma. In: *Advances in Neurosurgery* 20. Piscol, K.; Springer, M.; Brock, M., eds., Springer Verlag, Berlin-Heidelberg-New York, in the press (1991).
5. Slatkin, D.; Micca, P.; Forman, A.; Gabel, D.; Wielopolski, L.; Fairchild, R. Boron uptake in melanoma, cerebrum and blood from $Na_2B_{12}H_{11}SH$ and $Na_4B_{24}H_{22}S2$ administered to mice. *Biochem. Pharmacol.* 35, 1771-1776 (1986).
6. Fankhauser, H.; Stragliotto, G. Borocaptate sodium (BSH) pharmacokinetics in glioma patients. In: *Boron Neutron Capture Therapy: Toward Clinical Trials of Glioma Treatment.* Gabel, D.; Moss, R.L., eds., Plenum Press New York, 155 - 162 (1992).
7. Kraft, S.; Gavin, P.; DeHaan, C.; Leathers, C.W.; Bauer, W.; Dorn III, R. Biodistribution of boron in canine with spontaneous intracranial tumors following borocaptate sodium infusion. In: *Progress in Neutron Capture Therapy for Cancer.* Allen, B.J.; Moore, D.E.; Harrington, B.V., eds., Plenum Press, New York, 537-540 (1992).
8. Siefert, A.; Casado, J.; Moss, R.L.; Gavin, P.; Philipp, K.; Huiskamp, R.; Dühmke, E. Healthy tissue tolerance studies for BNCT at the high flux reactor in Petten - first results. In: *Boron Neutron Capture Therapy: Toward Clinical Trials of Glioma Treatment.* Gabel, D.; Moss, R.L., eds., Plenum Press, New York, 179 - 188 (1992).

MACROSCOPIC AND MICROSCOPIC BIODISTRIBUTION OF

BSH IN A RAT GLIOMA MODEL

Crister P. Ceberg[a], Anders Persson[d], Arne Brun[b],
Bengt Jergil[d], Bertil R.R. Persson[a], and Leif G. Salford[c]

[a]Dept of Radiation Physics
[b]Dept of Neuropathology
[c]Dept of Neurosurgery
Lund University Hospital
[d]Dept of Biochemistry
University of Lund
S-221 85 Lund, Sweden

INTRODUCTION

Boron Neutron Capture Therapy (BNCT) may have a unique capacity to cure astrocytomas if the distribution of the injected boron compound corresponds to the growth of the tumour. Astrocytomas grade III-IV grow in an octopus-like fashion, with arms of proliferating tumour tissue reaching along the tracts of white matter into the normal brain. An even more crucial fact is that isolated malignant cells migrate from the tumour into the surrounding brain where they hide like "guerillas" among normal cells behind an intact blood-brain barrier (BBB). These cells give rise to the hitherto unavoidable recurrencies - even hemispherectomies fail to radically remove the tumour[1].

Thus, a fully successful BNCT relies on that the boron compound is capable of reaching and penetrating the migrated tumour cells. For any boron compound aimed for BNCT, detailed boron distribution measurements, both on a macroscopic scale and on a microscopic scale, are therefore desirable.

Our group participates in the European Collaboration on BNCT, and has concentrated on studies of pharmacokinetics and biodistribution of BSH in a rat glioma model. In the present work, an improved method for Neutron Capture Radiography has been used to produce boron-images of tissue slices that can be compared with histological examinations of the same tissue slice. This gives information about the boron distribution relative to the spread of the tumour cells on a macroscopic scale.

Boron Neutron Capture Therapy, Edited by D. Gable
and R. Moss, Plenum Press, New York, 1992

The boron localization on a microscopic level has earlier been studied in cultured tumour cells by measurements on separated subcellular fractions e.g. by Nguyen et al[2]. In this work we have gone one step further with subcellular fractioning of cells from tumour bearing rats.

MATERIALS AND METHODS

An animal model with rats carrying a cerebral RG2 rat glioma that mimics very well the human situation was used in the experiments. Following a BSH injection corresponding to 25 μg ^{10}B/g body weight, the animals were sacrificed 12 hours later.

Macroscopic Measurement

Tissues of interest were excised, and frozen sections were prepared for Neutron Capture Imaging as described elsewere[3]. For quantitative analyses, the different shrinkage of different tissues during the freeze-drying process, was taken into account by using experimentally obtained correction factors[3]. The tissue sections were subsequently stained with methylene blue and used for comparison between the boron distribution and the histology.

Microscopic Measurement

The tumours from ten rats were excised and placed in ice-chilled 0.4M sucrose, 5mM Tris/HCl (pH=7.5). The material amounted to 1 g. After homogenization in a dounce the homogenate was ultracentrifugated at 100,000xg for 90 minutes. The pellet and the supernatant were collected and the boron concentration was measured with quantitative Neutron Capture Imaging.

RESULTS

Macroscopic Distribution

Figure 1 shows the histology and the corresponding boron distribution image of one slice through the tumor in a rat brain 6 hours after the BSH administration. In such images it has been found that BSH can reach small remote glioma metastases were the blood-brain barrier theoretically may be still intact[4]. Also, a spread of boron from the tumour into the surrounding edematous normal brain, has been found to arise with time[4].

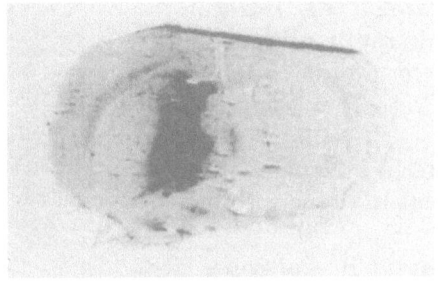 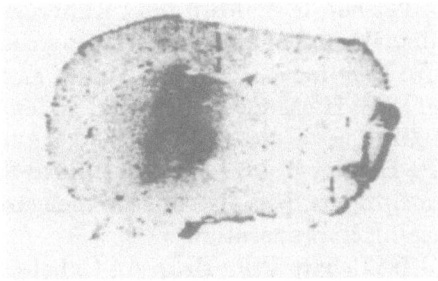

Figure 1. *Histology (left) and the boron distribution image (right) of a section from a rat brain.*

Microscopic Distribution

The ^{10}B content in the cell homogenate (3 ml) was determined to 3.0 μg, and to 0.8 μg/g and 2.1 μg/g in the pellet supension fraction (1 ml) and the supernatant fraction (1 ml) respectively. Thus, the percentage distribution of BSH was 30% in the pellet and 70% in the cytosol.

DISCUSSION

We have found that BSH can be revealed in small distant metastases, which supports the potential of BNCT to cure astrocytomas. In our continued research we strive to clarify whether or not boron can be brought to and accumulated even in individual migrating glioma cells, in spite of an intact blood-brain barrier.

The leakage of boron into the healthy brain, and the absorbed dose it might cause to the normal tissue, is a problem that must be taken into account in future dosimetry work.

The fact that 30% of the BSH remained in the pellet after the homogenization/centrifugation procedure, indicates that there is a binding of BSH to the cell membranes. These results are, however, preliminary as the concentrations were very small and difficult to measure with available techniques. Further studies are under way with higher BSH doses, and with more detailed fractionation schedules.

REFERENCES

1. Salford L.G.; Brun A.; Nirfalk S. Ten year survival among patients with supratentorial astrocytoma grade III and IV. *J. Neurosurg.* 69, 506-509 (1988).

2. Nguyen T.; Teicher, B.A.; Miura M.; Kahl S.B.; Brownell G.L. Intracellular distribution of various boron compounds in rat 9L gliosarcoma cells. *In: Progress in Neutron Capture Therapy for Cancer.* Allen, B.J.; Moore, D.E.; Harrington, B.V., eds., Plenum Press, New York, 381-386 (1992).
3. Ceberg C.P.; Hemler R.J.B.; Persson B.R.R.; Salford L.G. Neutron Capture Imaging for 10B-distribution measurements in tissue specimens. Manuscript in preparation.
4. Hemler R.J.B.; Ceberg C.P.; Brun A.; Gabel D.; Larsson B.; Persson B.R.R.; Salford L.G. A Quantitative Study on pharmacokinetics and biodistribution of BSH in a rat glioma model. In: *Progress in Neutron Capture Therapy for Cancer.* Allen, B.J.; Moore, D.E.; Harrington, B.V., eds., Plenum Press, New York, 469-474 (1992).

HEALTHY TISSUE TOLERANCE STUDIES FOR BNCT

AT THE HIGH FLUX REACTOR IN PETTEN -

FIRST RESULTS

Axel Siefert[a,c], Juan Casado[a], Raymond L. Moss[a],
Patrick Gavin[a,d], Katharina Philipp[b],
René Huiskamp[b], and Eckhart Dühmke[c]

[a]Commission of the European Communities
Joint Research Centre (JRC)
NL-1755 ZG Petten, The Netherlands

[b]Radiobiology and Radioecology
Netherlands Energy Research Foundation(ECN)
NL-1755 ZG Petten, The Netherlands

[c]Dept. of Radiotherapy
University of Göttingen
D(W)-3400 Göttingen, F. R. Germany

[d]Dept. of Veterinary Medicine
Washington State University (WSU)
Pullman, WA 99164-6610, USA

INTRODUCTION

Boron Neutron Capture Therapy (BNCT) offers a potential form of therapy for highly malignant brain tumours such as glioblastoma multiforme[1,2]. The High Flux Reactor of the Commission of the European Communities in Petten, The Netherlands, will serve as the neutron source for this treatment[3]. Before first clinical trials can be performed on patients, the tolerance of healthy tissue exposed to the radiation field has to be determined. Due to the characteristics of epithermal neutrons a large animal model has to be used for this task. Since the canine model has proved to be an excellent tool to answer this issue[4], it has been chosen for the healthy tissue tolerance studies at Pet-

Boron Neutron Capture Therapy, Edited by D. Gable
and R. Moss, Plenum Press, New York, 1992

ten. The paper here describes the experimental set up and the initial results from the first two canine experiments which were irradiated in July 1991.

MATERIALS AND METHODS

For the experiment adult female beagle dogs, age 12-24 months, with a weight ranging from 9-15 kg were obtained from a commercial animal supplier. They are housed at the "Gemeenschappelijk Diereninstituut Amsterdam" (GDIA).

Prior to the study extensive examinations and evaluations of the healthy state of the dogs were performed. The tests included: complete physical and neurological examination, complete blood count, serum biochemistries, urine analysis, cerebral spinal fluid analysis, and magnetic resonance imaging (MRI) of the head. This examination was repeated 24 hours and 2 weeks after irradiation and will be carried out 26 and 52 weeks post irradiation and whenever it is clinically indicated. The dogs are observed daily. The observation period, post irradiation, is scheduled to be at least 1 year. Some dogs will be observed for two years. The endpoints of the study are the occurrence of unacceptable effects such as neurological symptoms (e.g. seizures, paralysis) or severe dermal necrosis. Extensive histological evaluation of all organs of interest will be carried out afterwards.

For MRI a 4.7 tesla magnet (SISCO 200), operating at 200 MHz for proton imaging, is used (National in Vivo NMR Facility, Bijvoet Center, Utrecht). The images are obtained with a spin-echo-sequence in multislice technique (slice thickness 5 mm, gap 0 mm). Moderately T_2 weighted images are created with an echo time (TE) of 60 ms and a repetition time (TR) of 1000 ms. Moderately T_1 weighted images (TE 30 ms, TR 800 ms) are obtained before and after the administration of the contrast agent Gd-DTPA (Magnevist™, Schering AG, 0.1 mmol/kg body weight).

In our study, experiments will be performed at two different mean blood-^{10}B concentrations (25 and 50 mg ^{10}B/kg body weight). At each boron level three dosage groups (described by the peak physical dose) will be investigated. Five dogs will be irradiated at each dose point (Table 1). In addition, fractionated experiments are planned at the lower boron concentration, where the total dose will be divided into four fractions, separated by 24 hours. Due to the complexity of the mixed field irradiation, the total dose will not be increased for fractionation. Boron administration will be repeated prior to each fraction.

For our experiments the boron containing compound borocaptate sodium (BSH) (95% ^{10}B enriched, Centronic Ltd.) is used. The compound has been checked for mercury contamination (<10 ppm). An amount of BSH equivalent to 25 or 50 mg ^{10}B/kg b.w. is diluted in 11 ml 0.9% NaCl/kg b.w. and infused with an infusion pump over 60 minutes via a cephalic vein catheter. After the end of the infusion blood samples are collected through a jugular catheter. Boron content of these blood samples is analyzed with two different methods: i) Prompt-Gamma-Ray-Spectroscopy (PGS), and ii) Inductively Coupled Plasma-Atomic Emission Spectroscopy (ICP-AES). With PGS the sample can be analyzed without too much time consuming sample preparation. The PGS facility is conveniently situated on a neighbouring beam tube next to the irradiation facility. The results are available within 10-15 minutes after collecting the sample. This facility allows an adjustment of the total irradiation time according to the actual blood-boron concentration for each individual animal.

Table 1

Programme of experiments for the healthy tissue tolerance study on the canine brain

Single fraction

mean blood ^{10}B conc. [ppm]	Peak physical dose to blood(2.25cm)[Gy]	number of dogs
	15	5
25	19	5
	23	5
	23	5
50	27	5
	31	5

Four fractions

	15	5
25	19	5
	23	5

During irradiation dogs are kept under general anaesthesia. Introduction of anaesthesia is done i.v. with Acepromacine and Methadone. For maintenance, inhalation anaesthesia (Isoflurane in O_2/N_2O, controlled ventilation) is applied with a respirator. Prior to irradiation Dexamethasone (1 mg/kg b.w.) is injected i.v. Animals are monitored using an ECG.

Dosimeters (TLDs and activation foils) are attached to the surface of the dog at the following locations:

1: top of the head (christa saggitalis externa, centre of the cranium)
2&3: above the left and right eye (regio supraorbitalis)
4: rear of the skull (christa saggitalis externa, caudal end)
5: thyroid gland
6: oral cavity (attached to the respiration tube)
7: on the chest (7th intercostal space, regio cartilago costalis)
8: between the shoulders (regio interscapularis). More details can be found in this and other proceedings[5,6].

The physical characterisation of the epithermal neutron beam is described elsewhere[7]. During irradiation, the incident neutron beam used a circular beam delimiter of 8 cm diameter centred over the cranial vault while the dog was in left lateral recumbency.

To determine the blood-^{10}B concentration, blood samples are obtained before irradiation and after half of the projected total irradiation time. Analysis of these samples at the PGS facility makes it possible to calculate the mean blood ^{10}B concentration and adapt the total irradiation time according to the results. After the end of irradiation, an additional sample is obtained to verify the calculated mean blood ^{10}B concentration.

RESULTS

The first two dogs of this study were irradiated in July 1991. Both dogs were scheduled for the 50 ppm boron, 23 Gy group. Due to an unexpected

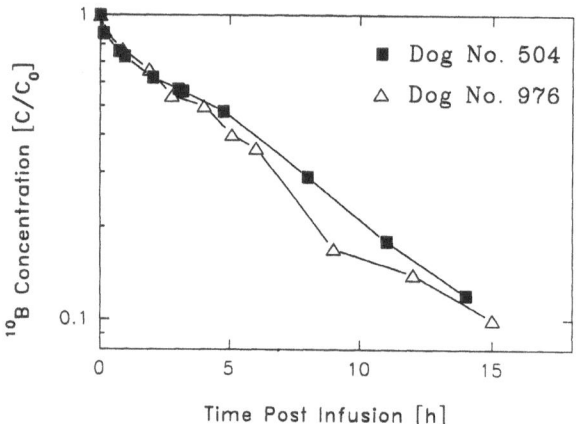

Figure 1. *Pharmacokinetic Profile for Blood ^{10}B concentration measured with ICP-AES.*

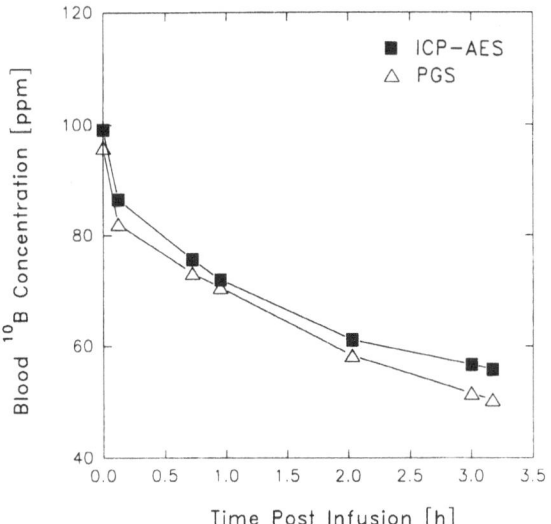

Figure 2. *Blood* ^{10}B *Concentration: Comparison ICP-AES vs. PGS backg =*
all reactions excluding the ^{10}B *component.*

high water content (30%) of the BSH used in this experiment, the first dog
(No. 976) received only 35 mg ^{10}B/kg b.w. with the consequence that the
mean blood ^{10}B concentration was only 38.5 ppm during irradiation. Irradiation time however was not changed (1.77 h) which resulted in a total physical
dose at peak of only 17.5 Gy.

The second dog (No.504) received a higher amount of BSH so that the
mean blood ^{10}B concentration during irradiation was 62.5 ppm. Total physical dose at peak was 24.5 Gy.

The BSH infusion was tolerated well by both dogs. Results of the pharmacokinetic analysis of the two dogs are shown in Fig. 1. The calculated half-life
for boron in the first four hours was 3.8 h for dog No. 976 and 4.9 h for dog
No. 504.

In Fig. 2, a comparison is shown of the results between PGS and ICP-AES.
Both methods are in good agreement, but the values obtained with PGS are
5-10% below those of the ICP-AES.

Dosimetry

In Table 2, the calculated physical doses are given for the two dogs at the
surface and at the peak dose position. The calculated depth dose distribution
profiles of the various dose components for dog No. 504 are shown in Fig. 3.

Table 2

Calculated physical doses at surface
and depth for the two irradiate dogs

animal No.	^{10}B conc. [ppm]	irrad. time [hrs.]	surface dose (0.25 cm) [Gy]		peak dose (2.25 cm) [Gy]	
			^{10}B	4.30	^{10}B	12.02
			fast n	2.29	fast n	0.96
			gamma	2.75	gamma	3.66
976	38.5	1.77	N(n,p)	0.19	N(n,p)	0.55
			other	0.11	other	0.31
			total	9.64	total	17.50
			^{10}B	6.94	^{10}B	19.42
			fast n	2.29	fast n	0.96
			gamma	2.75	gamma	3.66
504	62.3	1.77	N(n,p)	0.19	N(n,p)	0.55
			other	0.11	other	0.31
			total	12.28	total	24.90

Blood Analysis

Both dogs showed a depression of the platelet count after two weeks. The values of dog No. 976 went down to $82.5 \cdot 10^9/l$, those of No. 504 decreased to $51 \cdot 10^9/l$ (normal range for dogs: $150 - 400 \times 10^9/l$). These values were back to normal when analyzed one week later. A minor decrease of the total white blood count was observed two weeks after irradiation in dog No. 504. The other blood parameters did not show any significant changes after irradiation.

Magnetic Resonance Imaging

Two MRIs per dog were carried out after irradiation (24 h and 14 days post irradiation). With our technique using Gd-DTPA, no leakage of the blood brain barrier (BBB) was detected and the MRIs were normal at 24 h and 2 weeks post irradiation.

Physical and Neurological Examination

Beside skin reactions none of the dogs showed abnormalities.

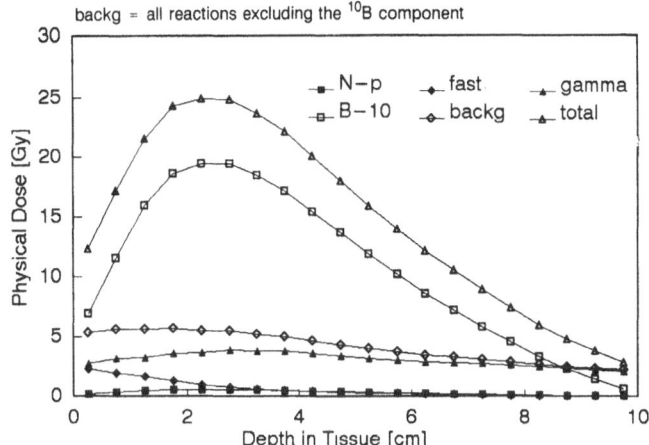

Figure 3. *Depth Dose Distribution, Dog No. 504.*

Skin

Dog No. 976. Initial epilation occurred 22 days after irradiation. Multiple areas of 2-3 mm in size were present in irradiated skin where hairs could be pulled out easily. Within the following two weeks the lesions increased to a circular field of 10 cm diameter. Several dozen hyperpigmented spots (size <2 mm) developed in the irradiated skin. Moist desquamation or skin necrosis was not present. Eight weeks after irradiation the hair started to regrow with reduced pigmentation.

Dog No. 504. Initial epilation occurred 19 days after irradiation and became total to the maximum size of 10.5 cm. Some "islands" in the rostral region of the irradiation field did not lose hair. Again, no moist desquamation or skin necrosis was present. Hyperpigmentation of the skin did not occur in this dog. Hair regrowth started 7 weeks after irradiation with reduced pigmentation.

DISCUSSION

It should be stressed that the results reported above are preliminary and an additional 43 dogs are still to be irradiated.

Both dogs were scheduled for the 23 Gy dose group at 50 ppm mean blood ^{10}B concentration. Due to the high water content of the used BSH, the aimed mean blood ^{10}B concentration could not be reached in dog No. 976. Because of concern about the high skin dose (fast neutrons) resulting from prolonged irradiation the pre-calculated irradiation time was not increased.

The discrepancy between the intended 23 Gy and the applied 24.9 Gy for dog No. 504 was caused by the underestimation of the blood ^{10}B concentration by the PGS compared to ICP-AES (5-10%). This fact will be accounted for in future irradiations.

The 24 h MRI is performed in order to check for acute irradiation changes, especially a break down of the BBB. The integrity of the BBB with respect to fractionated irradiation is of great importance[8]. Repeated BSH administration could increase the 10B content in areas of the brain where the BBB is damaged and thus lead to a higher radiation dose in these areas. The fact that no visible damage of the BBB could be detected in the 24 h MRI indicates that fractionated therapy may possibly be performed in that time frame.

In both dogs the platelet count decreased slightly after two weeks. Although this was of no clinical concern for the two irradiated dogs, the depression indicates the total body dose. Platelet depression of a greater magnitude was seen in beagle dogs following a total body dose of 2.4 Gy (x-rays)[9]. CBC analysis will be performed more frequently in future on our dogs to obtain more information concerning these findings.

The irradiation time was identical for both dogs (1.77 h). Therefore the only dose component that differed was the ^{10}B$(n,\alpha)^7$Li dose. Although a difference of 2.65 Gy physical dose at surface was calculated, the skin reactions of both dogs were almost identical, with the exception of a small time difference in the onset of the first skin changes. Since this difference was also present in the beginning of hair regrowth, it could simply be caused by biological variability. However, higher doses are often associated with an earlier onset of clinical changes. The fact that no significant difference in the skin reactions is present, indicates a compound factor below 1 for the ^{10}B$(n,\alpha)^7$Li reaction in the dog skin. This is in good agreement with the findings of Gavin et al.[10], who also suggested compound factors for the ^{10}B reaction below 1.

The skin changes of the two dogs irradiated in Petten are similar to those observed at Washington State University (WSU)[10]. Their dogs were irradiated with the epithermal beam of the Brookhaven Medical Research Reactor (BMRR) after the infusion of similar amounts of BSH. Skin changes in the 27 Gy group were limited to epilation and dry desquamation. Moist desquamation and dermal necrosis did not occur[10]. This group may be equivalent to 25 Gy of the Petten beam. However, some of the WSU dogs developed changes in the brain that were visible as cortical contrast enhanced areas in the MRI after 4-6 months, but were not accompanied by neurological symptoms. Dog No. 504 is at a greater risk for developing such changes[10]. Additional MRIs will be performed 4 months after the irradiation to monitor these potential changes. If such lesions should occur at this dose level, the highest dose group (31 Gy physical dose at peak, 50 ppm ^{10}B) might exceed brain tolerance and the irradiation protocol may need modifying.

ACKNOWLEDGMENTS

We are grateful to Schering AG, Berlin, for a generous gift of Magnevist. Part of this research was carried out at the National in vivo NMR Facility at the Bijvoet Center for Biomolecular Research of the University of Utrecht. This study has been financially supported by the Commission of the European Communities.

REFERENCES

1. Barth, R.F.; Soloway, A.H.; Fairchild, R.G. Boron neutron capture therapy of cancer. *Cancer Res.* 50, 1061 (1990).
2. Hatanaka, H., ed. *Boron Neutron Capture Therapy for Tumors.* Nishimura Co. Ltd., Asahimachi-dori, Niigata, Japan (1986).
3. Moss, R.L.; Stecher-Rasmussen, F.; Ravensberg, K.; Constantine, G., Watkins, P. Design, construction and installation of an epithermal neutron beam for BNCT at the high flux reactor, Petten. In: *Progress in Neutron Capture Therapy for Cancer.* Allen, B.J.; Moore, D.E.; Harrington, B.V., eds., Plenum Press, New York, 63-66 (1992).
4. Gavin, P.R.; Kraft, S.L.; DeHaan, C.E.; Griebenow, M.L.; Moore, M.P. A large animal model for boron neutron capture therapy. In: *Progress in Neutron Capture Therapy for Cancer.* Allen, B.J.; Moore, D.E.; Harrington, B.V., eds., Plenum Press, New York, 479-484 (1992).
5. Stecher-Rasmussen, F.; Constantine, G.; Freudenreich, W.; de Haas, H., Moss, R.L.; Paardekooper, A.; Ravensberg, K.; Verhagen, H.; Voorbraak, W.; Watkins, P.R.D. From filter installation to beam characterization. In: *Boron Neutron Capture Therapy: Toward Clinical Trials of Glioma Treatment.* Gabel, D.; Moss, R.L., eds., Plenum Press, New York, 59 - 77 (1992).
6. Moss, R.L.; Siefert, A.; Watkins, P.R.D.; Constantine, G.; Stecher-Rasmussen, F.; Philipp, K. I - Phantom dosimetry techniques used in the nuclear and biological characterisation of the epithermal neutron beam and in the healthy tissue tolerance studies on the canine brain, and II - The development towards a treatment planning system for the treatment of glioma patients by BNCT. *Int. Workshop on Macro and Microdosimetry and Treatment Planning for Neutron Capture Therapy.* Plenum Press, New York, in the press (1992).
7. Watkins, P.; Konijnenberg, M.; Constantine, G.; Rief, H.; Ricchena, R.; de Haas, J.B.M.; Freudenreich, W. Review of the physics calculations performed for the BNCT facility at the HFR Petten. In: *Boron Neutron Capture Therapy: Toward Clinical Trials of Glioma Treatment.* Gabel, D.; Moss, R.L., eds., Plenum Press, New York, 47 - 58 (1992).
8. Dorn III, R.V.; Spickard, J.H.; Griebenow, M.L. The effects of ionizing radiation and dexamethasone on the blood-brain-barrier and blood-

tumor-barrier: Implications for boron neutron capture therapy for brain tumors. *Strahlenther. Oncol.* 165, 219 (1989).

9.   Baltschukat, K.; Nothdurft, W. Hematological effects of unilateral and bilateral exposures of dogs to 300-kVp x-rays. *Radiat. Res.* 123, 7 (1990).

10. Gavin, P.R.; Wheeler, F.J.; Huiskamp, R.; Siefert, A.; Kraft, S.; DeHaan, C. Large animal model studies of normal tissue tolerance using an epithermal neutron beam and borocaptate sodium. In: *Boron Neutron Capture Therapy: Toward Clinical Trials of Glioma Treatment.* Gabel, D.; Moss, R.L., eds., Plenum Press, New York, 197 - 209 (1992).

CELLULAR PHARMACOKINETICS OF BNCT COMPOUNDS

AND THEIR CELLULAR LOCALIZATION WITH EELS/ESI

Ruud Verrijk[a], René Huiskamp[b], I.J.H. Smolders[a],
Adrian C. Begg[a], C.W.J. Sorber[c], and W.C. De Bruijn[c]

[a]Division of Experimental Therapy
The Netherlands Cancer Institute
NL-1066 CX Amsterdam, The Netherlands

[b]Department of Radiobiology and Radioecology
Netherlands Energy Research Foundation (ECN)
NL-1755 ZG Petten, The Netherlands

[c]AEM Unit
Clinical Pathological Institute I
Erasmus University
NL-3075 EA Rotterdam, The Netherlands

INTRODUCTION

The principle of Boron Neutron Capture Therapy is based on the nuclear reaction that occurs when ^{10}B, a stable isotope, is irradiated with thermal neutrons (0.025 eV neutrons). Particles, emitted from the capture reaction are largely high LET α-particles and lithium-7 particles with a mean free path of 9 μm and 5 μm, respectively. If tumor cells can be selectively loaded with boron containing compounds, unresectable tumors may be eradicated with a high normal tissue tolerance by irradiating the tumor volume with thermalized neutrons from an epithermal (10 keV mean energy) neutron beam. The primary goal in current BNCT research is to examine the therapeutic potential of BNCT for the treatment of grade IV gliomas, although successive objectives include treatment feasibility studies of other tumors, with a main interest in melanomas[1,2].

Boron Neutron Capture Therapy, Edited by D. Gable
and R. Moss, Plenum Press, New York, 1992

The microdosimetry of high LET particles resulting from boron neutron capture reactions depends on tumor boron concentration, intratumoral (intercellular) boron distribution and intracellular boron distribution. The RBE for the neutron capture recoil particles is for a great deal dictated by these parameters which will influence the therapeutic efficacy and normal tissue tolerance. Monte Carlo simulations of the dose distribution of emitted α-particles and ^{7}Li-particles in tumor tissue have estimated a minimum boron concentration of 20 μg/g tissue for an effective treatment if boron is distributed homogeneously throughout the cell. For a hypothetical cubic cell with representative dimensions (1150 μm^3 and a spherical nucleus of 230 μm^3) a 7-fold reduction in efficacy was found for exclusively extracellularly located boron, and a 5-fold increase for intranuclearly located boron compared with a homogeneous boron distribution[3]. Another requirement is to achieve an acceptable normal tissue tolerance by aiming at preferential uptake and retention of boron in tumor cells versus normal tissue cells.

Total cell boron is an important factor in BNCT dosimetry. We therefore investigated the in vitro cellular uptake and retention of borocaptate sodium (BSH) and 5-dihydroxyboryl-6-propyl-2-thiouracil (BPTU-1) in B16 murine melanotic melanoma cells. Uptake of BSH was also measured in rat embryo fibroblasts, which served as a normal skin tissue equivalent cell line. Analysis of boron in cell pellets and cell culture media was done with Inductively Coupled Plasma Atomic Emission Spectroscopy (ICP-AES).

In addition to total cell uptake, we also wanted information on subcellular drug distribution. Quantitative Neutron Capture Radiography (QNCR, Track Etch) has been used to measure intratumoral boron distributions, but due to the physical principle of this method, the spatial resolution is limited to approximately 5-10 microns, hence delivering insufficient information about boron distribution at a subcellular level[4]. Secondary Ion Mass Spectroscopy has been used for boron imaging, but this technique also lacks a spatial resolution below one micron[5]. In an attempt to elucidate the intracellular localization of boron in cells or tissue, a study was initiated to establish the feasibility of the application of Electron Energy Loss Spectroscopy (EELS) and Electron Spectroscopic Imaging (ESI) for intracellular boron distribution images.

EELS is based on energy loss of beam electrons by element- specific ionization. For boron, the absorption edge in the energy loss spectrum is $\Delta E = 188$ eV.

ESI is theoretically capable of detecting down to 3000 boron atoms with a spatial resolution of 0.5 nanometer[6,7]. Normally, endogenous boron concentrations are extremely low, which offers a low background signal that translates into a high sensitivity.

Using EELS and ESI, we have looked at boron distribution in Epon embedded B16 melanoma cells after incubation with BSH in vitro.

METHODS

In Vitro Cellular Boron Uptake

Melanotic B16 murine melanoma cells and rat embryo fibroblasts were grown in Dulbecco's modified Eagle medium containing 10% Foetal Calf Serum in an incubator with 5% CO_2 at 37°C. Cells were trypsinized and kept in suspension by gently rotating the vial. The cells were exposed to BSH (1, 3, 10 and 30 $\mu g/g$ B in cell culture medium) or BPTU-1 (1, 3 and 10 $\mu g/g$ B) for several periods of time up to 16 hours. In boron retention experiments, cells were washed for several periods of time in boron-free culture medium after they had been preincubated with the boron compound. For bulk boron analysis cells were pelleted by centrifugation and the boron containing medium was removed. Medium residues on top of the pellet and sticking to the vial wall were removed as much as possible with a cotton tip. This way, a reproducible 13-15% contamination of the pellet with medium was obtained. Samples were spiked with a known amount of cobalt nitrate as recovery standard. Boron and cobalt in cell pellets and culture media were analyzed by ICP-AES after acid digestion. The detection limit of the ICP-AES was established to be approximately 0.01 $\mu g/g$ B, which allowed us to measure in the sub-ppm range in cell material. Pellet to medium boron ratios above unity were considered indicative of cellular accumulation of boron compounds.

Electron Energy Loss Spectroscopy

Energy loss spectra and ESIs were taken on a Zeiss EM902 electron microscope equipped with an EELS spectrometer unit, a photomultiplier tube and an IBAS 2000 video image analysis system. Electron spectroscopic images were acquired at pre-ionization and ionization edges of 168 and 208 eV, respectively. A boron spectrum was taken from BSH, precipitated onto a membrane furnished copper grid. B16 cells pre-incubated for 16 hours with 200 $\mu g/g$ boron as BSH were used for Electron Spectroscopic Imaging. Fixation was accomplished in 1 hour by adding glutaraldehyde to the cell culture medium to a final concentration of 2% (v/v). Cells were embedded in Epon after dehydration with a five-step water/acetone gradient. Ultrathin sections of 40-80 nm were cut on an LKB ultramicrotome and transferred onto a copper grid. Net boron distribution images were acquired with the extrapolated background subtracted.

RESULTS

Drug Uptake

Uptake of BSH (1, 3, 10, and 30 $\mu g/g$ boron) in B16 showed preferential accumulation in the cell pellet only after 16 hours incubation (Fig. 1, panel A).

The ratio pellet/medium was above 3 for all measured incubation concentrations. After 10 and 30 μg/g medium boron incubations boron levels of greater than 30 μg/g in the cell pellet were found. In rat embryo fibroblasts, the accumulation ratio was 3 for 1 μg/g B incubation, but it declined with increasing medium boron concentration to a value of 1 (Fig. 1, panel B).

In a separate experiment, 15 hours pre-incubation of B16 cells without BSH and a subsequent 1 hour incubation with BSH revealed that co-incubation of cells and BSH for 16 hours was essential to achieve this accumulation (data not shown).

Accumulation ratios of BPTU-1 in B16 ranged from 1.2 to 5, and were inversely related to the boron medium concentration (Fig. 1, panel C).

Such a relationship may imply a saturable uptake process in the cell. After 16 hours of incubation with 10 and 30 μg/g B as BSH, B16 cells lost

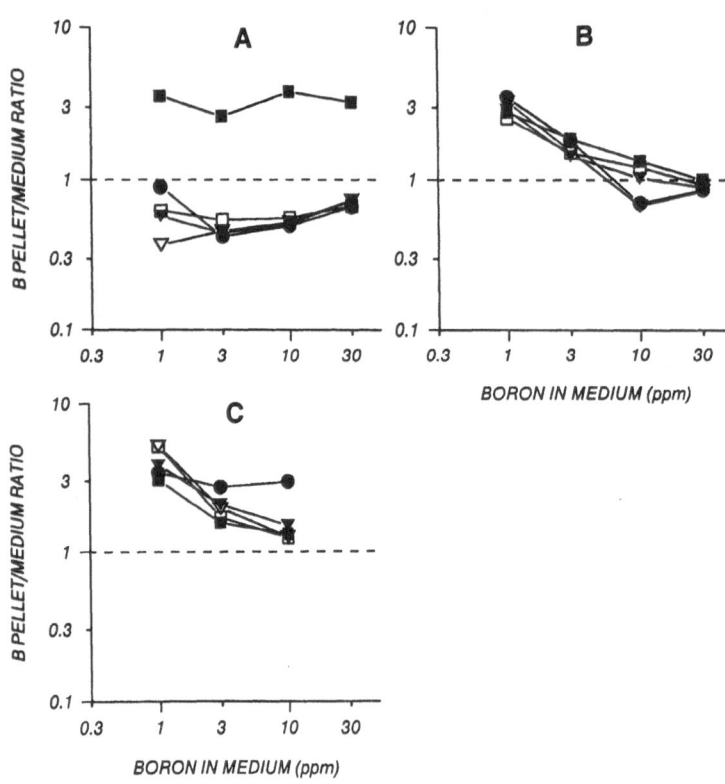

Figure 1. *Boron accumulation ratios after incubations of 30 min (•), 1 hour (▽), 2 hours (▼), 4 hours (□) and 16 hours (■) of BSH with B16 melanoma cells (A), BSH with rat embryo fibroblasts (B), and BPTU-1 with B16 cells (C).*

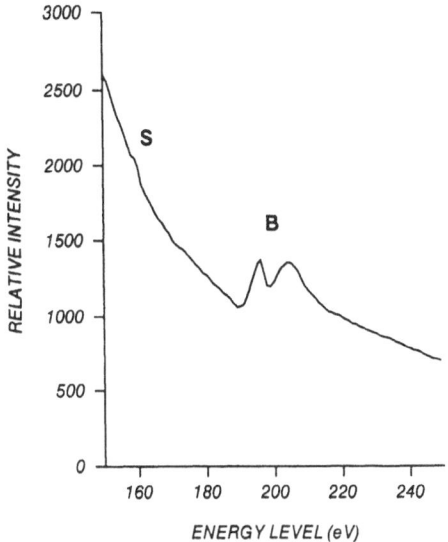

Figure 2. *Energy loss spectrum of BSH on a Formvar film. B denotes the boron ionization edge, S indicates the ionization edge of sulfur.*

most of their accumulated boron during the first 15 minutes, but boron levels were maintained between 4 and 5 µg/g during at least 2 hours washing in boron-free medium. Cells preloaded with 1 and 3 µg/g B as BPTU-1 showed constant retention of boron (5 µg boron per gram cell pellet) after 15, 45 and 120 minutes of re-incubation of B16 cells in boron-free culture medium.

Figure 3. *Electron spectroscopic boron image of a cytoplasmic region in B16 melanoma cells, containing BSH in cellular compartments.*

EELS and ESI Analysis

In the EELS spectrum, the boron ionization edge of BSH, placed on a Formvar film equipped grid, was detected at 188 eV in the EELS spectrum after extrapolated background subtraction. The residual energy loss signal indicated a sensitivity which was considered sufficient for ESI analysis (Fig. 2).

ESI images of BSH-loaded B16 melanoma cells revealed that boron was present in subcellular compartments, while boron was undetectable in the nucleus and in the free cytosol (Fig 3). This indicates that the present sample fixation and preparation techniques are not yet optimal, since BSH must have been present in the cytosol to enter specific cell organelles. Fixation of the sulfhydryl moiety of BSH to other fixable material may have been slower than expected and cytosolic boron was probably lost during sample preparation. The demonstrated boron image presumably reflects the boron which may have been incorporated in the melanin polymer. Transmission electron microscopic images of the cells demonstrated that the compartments were probably melanosomes (not shown). Although no boron images of BPTU-1 have yet been produced, it is anticipated on theoretical grounds that after sufficiently long incubation periods melanoma cells will probably express a retained boron content in melanosomes in the form of boronated melanin.

DISCUSSION

It has been shown that owing to a defective blood-brain barrier in tumor vasculature a 'mechanical' selectivity of BSH for tumors in cranial areas exists[8]. The present results indicate that cellular accumulation of BSH, at least in B16 melanoma cells, may also have a biochemical rationale. At first sight, BSH is not a direct melanin precursor analog such as l-boronophenylalanine. However, it is known that during melanogenesis, specific cellular sulfhydryl compounds are transported across the cell membrane as building blocks for melanin, and that they are also of significant importance in the regulation of melanogenesis[9]. It is therefore possible that BSH parallels physiological sulfhydryl compounds in passing cell membranes and entering cell organelles through similar uptake systems.

Intracellular localization of BNCT compounds seems to be feasible, at least for BSH at the concentration chosen in this study in B16 melanoma cells. Melanoma treatment is certainly a field where BNCT has a potential, mostly due to the relatively high probability of development of tumor seeking drugs with a biochemical rationale, i.e. via the melanin synthesis pathway. BSH, however, the drug currently scheduled for clinical studies, has no obvious biochemical rationale for tumor uptake and it is therefore important to investigate boron distribution in glioma cell lines and in glioma tissue.

Current studies in our laboratory are focused on cellular uptake of other potential BNCT compounds and on adaptation of the EELS/ESI technique for the quantitative determination of subcellular boron distributions in cell lines and in tumor and normal tissues of animal and human origin.

ACKNOWLEDGMENT

This study was supported by grant NKI 90-10 from the Dutch Cancer Society.

REFERENCES

1. Hatanaka, H. Boron-neutron capture therapy for tumors. In: *Glioma*. Karim, A.B.M.F.; Laws, E.R., eds., Springer, Berlin (1991).
2. Coderre, J.A.; Kalef-Ezra, J.A.; Fairchild, R.G.; Micca, P.L.; Reinstein, L.E.; Glass, J.D. Boron neutron capture therapy of a murine melanoma. *Cancer Res.* 48, 6313-6316 (1988).
3. Gabel, D.; Foster, S.; Fairchild, R.G. The Monte Carlo simulation of the biological effect of the $^{10}B(n,\alpha)^7Li$ reaction in cells and tissue and its implication for boron neutron capture therapy. *Radiat. Res.* 111, 14-25 (1987).
4. Fairchild, R.G.; Gabel, D.; Laster, B.H.; Greenberg, D.; Kiszenick, W., Micca, P.L. Microanalytical tools for boron analysis using the $^{10}B(n,\alpha)^7Li$ reaction. *Med. Phys.* 13, 50-56 (1986).
5. Ausserer, W.A.; Ling, Y.; Chandra, S.; Morrison, G.H. Quantitative imaging of boron, calcium, magnesium, potassium, and sodium distributions in cultured cells with ion microscopy. *Anal. Chem.* 61, 2690-2695 (1989).
6. Bendayan, M.; Barth, R.F.; Gingras, D.; Londoño, I.; Robinson, P.T.; Alam, F.; Adams, D.M.; Mattiazzi, L. Electron spectroscopic imaging for high-resolution immunocytochemistry: Use of boronated protein A. *J. Histochem. Cytochem.* 37, 573-580 (1989).
7. Joy, D.C.; Maher, D.M. Electron energy loss spectroscopy: Detectable limits for elemental analysis. *Ultramicroscopy* 5, 333-342 (1980).
8. Hatanaka, H.; Moritani, M.; Camillo, M. Possible alteration of the blood-brain barrier by boron neutron capture therapy. *Acta Oncol.* 30, 375-378 (1991).
9. Aroca, Solano, F.; Martinez, J.H.; Lozano, J.A. The role of sulfhydryl compounds in mammalian melanogenesis: the effect of cysteine and glutathione upon tyrosinase and the intermediates of the pathway. *Biochim. Biophys. Acta* 967, 296-303 (1988).

LARGE ANIMAL MODEL STUDIES OF NORMAL TISSUE

TOLERANCE USING AN EPITHERMAL NEUTRON BEAM AND

BOROCAPTATE SODIUM

Patrick R. Gavin[a,d], Floyd J. Wheeler[b], René Huiskamp[c],
Axel Siefert[d,f], Susan Kraft[e], and Constance DeHaan[a]

[a]Washington State University
Pullman, WA 99164-6610, USA

[b]EG&G Idaho Inc.
Idaho Falls, ID 83415-3515, USA

[c]Netherlands Energy Research Foundation (ECN)
NL-1755 ZG Petten, The Netherlands

[d]Joint Research Centre
NL-1755 ZG Petten, The Netherlands

[e]Kansas State University
Manhattan, KS 66506-0112, USA

[f]University of Göttingen
D(W)-3400 Göttingen, F. R. Germany

INTRODUCTION

Epithermal neutron beams are being developed for potential boron neutron capture therapy (BNCT) to allow treatment of deep seated tumors, like glioblastoma multiforme, through the intact skin. The neutron capture cross-sections for elements in normal tissue are several orders of magnitude lower than that for boron but due to the relatively high concentrations of hydrogen and nitrogen in normal tissue, their capture through the $^1H(n,\gamma)^2H$ and the $^{14}N(n,p)^{14}C$ reactions respectively contribute significantly to the total radiation absorbed dose. Additional sources which contribute to the absorbed radiation dose are incident gamma-radiation and fast neutrons, i.e. components in the epithermal neutron beam.

A large animal model has to be used to study the tolerance of critical normal tissues to this form of radiotherapy using an epithermal neutron beam. In rodents and other small animals, the total body dose is intolerable and prevents studies on late effects. The total body irradiation is mainly caused by the gamma components coming from the $^1H(n,\gamma)^2H$ reaction in the irradiation field.

The pharmacokinetics of the boron compound borocaptate sodium ($Na_2B_{12}H_{11}SH$) hereafter called BSH have been studied extensively in the dog[1]. The knowledge of the macrodistribution of this compound in all normal tissues allows estimates of the radiation dose contribution of the boron neutron capture reaction to critical normal tissues[2]. Complete sampling of all normal tissues in the dog augments the limited sampling possible during human pharmacokinetic studies[3].

Large animal normal tissue tolerance studies play an important role in the design of patient irradiations using new radiation modalities[4-7]. Knowledge of tolerance of the superficial overlying tissues and most importantly of the brain have been determined prior to initiation of new radiation treatments for brain tumors[4,5].

The following study documents the initial studies of epithermal neutron irradiation and boron neutron capture using a canine model and BSH as the boron carrier.

MATERIALS AND METHODS

The pharmacokinetics of BSH was studied in normal laboratory dogs and in dogs with spontaneous occurring brain tumors. The amount of compound administered was varied from 20-100 mg boron/kg body weight. Sampling of the blood and body tissues was from 2-12 hours post administration. Detailed descriptions of the protocol have been previously reported[1]. The boron levels were determined by inductively coupled plasma-atomic emission spectroscopy (ICP-AES).

Twenty-five Labrador dogs have been irradiated with the epithermal neutron beam at the Brookhaven Medical Research Reactor (BMRR) Brookhaven, N.Y., USA, using a 5x10 cm portal and two Beagle dogs at the epithermal neutron beam at the High Flux Reactor (HFR) in Petten, The Netherlands, using a 8 cm diameter portal. Ten of the dogs irradiated at BMRR were with the epithermal neutron beam alone (EPI dogs), five dogs were irradiated to a dose of 11 Gy and five to 16.5 Gy. The other 13 dogs at BMRR and the 2 dogs at HFR were irradiated with epithermal neutron beams and blood boron concentrations of 31-62 μg boron/g blood (BORON dogs). The radiation dose varied from 0-64 Gy. The dosage groups were defined by the peak physical dose that occurred at 2-3 cm beneath the surface.

Table 1

a. *Early skin reactions*

Biological endpoint	Ref. X-ray dose (Gy)[1]	Radiation component	Phys. dose (Gy)	Eq. dose (Gy*RBE)[2]	Calculated fast neutron RBE[3]
Moist desquamation	18.3	N(n,p)	1.35	3.65	
		Gamma	4.49	4.49	
		Other	0.52	0.52	
		Fast n	3.18		3.0
Moist desquamation + ulceration	24.4	N(n,p)	1.35	3.65	
		Gamma	4.49	4.49	
		Other	0.52	0.52	
		Fast n	3.18		4.9

b. *Late skin reactions*

Biological endpoint	Ref. x-ray dose (Gy)[1]	Radiation component	Phys. dose (Gy)	Eq. dose (Gy*RBE)[2]	Calculated fast neutron RBE[3]
50% skin necrosis	36.6	N(n,p)	2.03	5.48	
		Gamma	6.76	6.76	
		Other	0.78	0.78	
		Fast n	4.78		4.9
>50% skin necrosis	>36.6	N(n,p)	2.03	5.48	
		Gamma	6.76	6.76	
		Other	0.78	0.78	
		Fast n	4.78		>4.9

Footnotes to Tables 1a and 1b

[1]Park et al., *Vet. Radiol.* 15, 108-111 (1974), ref. 10

[2]RBE Gamma and other = 1.0; RBE N(n,p) = 2.7
Archambeau, in: *Clinical Aspects of Neutron Capture Therapy.*
Plenum Press, New York, 9-20 (1989).

[3]$RBE_{fast\ neutrons}$ = {Ref. x-ray dose-[(Eq. dose gamma)+(Eq. dose N(n,p)) +(Eq. dose other)]}/Phys. dose fast neutron

The doses were adjusted to skin dose for the determination of skin RBE. Details of the dosages have been provided in earlier papers[8,9].

All dogs had serial physical and neurological examinations, complete blood counts, cerebrospinal fluid analysis, and brain imaging via computed tomography (CT) and/or magnetic resonance imaging (MRI). The brain imaging was performed prior and following the administration of appropriate contrast agents. The protocol details have been reported[8]. This report will limit the findings to the observed acute and late skin reactions and the changes seen on MRI. The observations in relation to doses used were compared to historical values reported in the literature for conventional photon irradiation in order to estimate the relative biological effectiveness (RBE) of the fast neutron (>10 keV) component of the epithermal beam component[4,10]. The complexity of the boron reaction involving microdosimetry has necessitated the term "compound factor" instead of RBE to describe the relative dose contribution of this component[11]. Radiation dose component calculations have been reported[2,12].

The RBE for fast neutrons was derived using the EPI dog tolerance studies by subtracting the physical doses multiplied by their RBE of the various beam components. The RBEs used in the calculations were equal to 1.0, 2.7 and 1.0 for all gamma sources, the $^{14}N(n,p)^{14}C$ reaction and other capture reactions respectively. The "compound factor" for the $^{10}B(n,\alpha)^7Li$ reaction was derived using a similar procedure as described for the calculation of the fast neutron RBE. The fast neutron RBE derived from the EPI dog tolerance

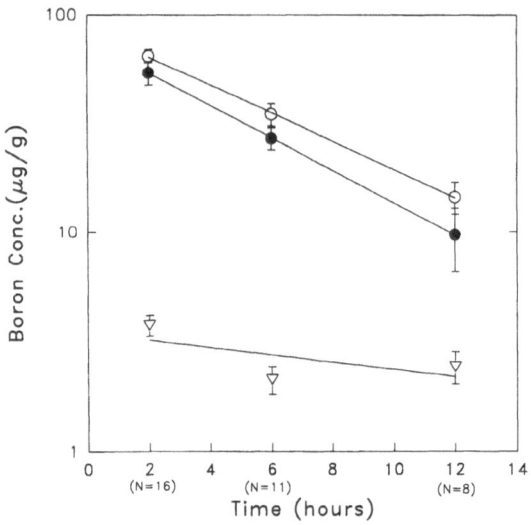

Figure 1. *Mean boron concentration (µg/g) in blood (O), skin of the head (•) and brain (▽) versus time after end of intravenous BSH infusion (55 mg $^{10}B/kg$).*

study was used as additional input leaving only the equivalent boron capture dose as variable. The value of 4.5 for the fast neutron RBE was chosen for these calculations.

RESULTS

The boron concentration in the skin of the head was slightly lower than the blood boron concentration. The brain parenchyma had significantly less boron than the blood. All concentrations decreased with time and the slopes of the decrease in the three tissue components were not statistically different, see Fig. 1.

The observed skin changes were biphasic with the early epidermal changes most prominent at 3 to 4 weeks and the dermal necrosis occurred at 12 weeks post-irradiation. Dry desquamation with small patchy areas of moist desquamation and epithelial ulceration occurred in the EPI dogs exposed to 11 Gy peak (3 cm depth) physical dose. Epidermal changes scored as moist desquamation with ulceration in greater than 50% of the incident radiation field, were observed in the 16.5 Gy EPI dogs and in the BORON dogs that received more than 38 Gy peak physical dose. The late skin reaction, i.e. dermal necrosis, occurred in the 16.5 Gy EPI dogs and in the BORON dogs that received 64 Gy peak physical dose. The 64 Gy BORON dogs were euthanized following dermal necrosis and several unsuccessful surgical attempts to alleviate the problem. The 16.4 Gy EPI dogs were euthanized following dermal necrosis.

The 11 Gy EPI dogs did not develop any visible CNS lesions within one year post-irradiation. The 39 Gy BORON dogs had fatal white matter necrosis of the cerebrum that appeared at 22 weeks post-irradiation. Two of three 27 Gy BORON dogs developed small contrast enhancing lesions of the cerebral white matter at six months post-irradiation. However, the 27 Gy BORON dogs were clinically normal at that time and remained so for the 12 months of the study. The contrast enhancing lesions were not present on the MRI examination 12 months post-irradiation. Figures of the representative reactions have been previously reported[8].

The fast neutron RBE for the acute and late skin effects were derived using published data for the other beam components and compared with skin tolerance doses determined for 200 kVp x-rays. Separate values for the acute and late skin effects were calculated as shown in Table 1.

The dose modification of the boron capture dose, or compound factor, was determined by using the published and above calculated RBEs and comparison of the remaining boron capture physical dose component with published X-ray and gamma tolerance doses for the skin and brain of the dog. Separate compound factors were derived for the skin and CNS, see Table 2. Using the derived RBE and compound factors, a summary of the experimental groups physical and equivalent dose is given in Table 3.

Table 2

Compound factor

Biological endpoint	Ref. x-ray dose (Gy)[1]	Radiation component	Phys. dose (Gy)	Eq. dose (Gy*RBE)[2]	Calculated compound factor[3]
Skin moist desquamation w & w/o ulceration	24.4		38 Gy dog group		
		N(n,p)	0.92	2.48	
		Gamma	3.06	3.06	
		Other	0.36	0.36	
		Fast n	2.17	9.77	
		B(n,α)	17.95		0.51
Dermal necrosis	36.6		64 Gy dog group		
		N(n,p)	1.24	3.35	
		Gamma	4.11	4.11	
		Other	0.48	0.48	
		Fast n	2.91	13.10	
		B(n,α)	29.90		0.52
CNS necrosis	14.8		27 Gy dog group		
		N(n,p)	0.68	1.84	
		Gamma	3.19	3.19	
		Other	0.42	0.42	
		Fast n	0.57	2.57	
		B(n,α)	21.58		0.32

[1]Skin effects: Park et al., *Vet. Radiol.* 15, 108-111 (1974), ref. 10
CNS effects: Fike et al., *Radiat. Res.* 99, 294-310 (1984), ref. 4
[2]RBE Gamma and other = 1.0; RBE N(n,p) = 2.7; RBE fast n = 4.5
[3]Compound factor$_{B(n,\alpha)}$={Ref. X-ray dose-[(Eq. dose gamma)+ (Eq. dose N(n,p))+(Eq. dose other)+(eq. dose fast n)]}/ Phys. dose B(n,α)

The two HFR irradiated dogs showed dry desquamation and epilation similar to that observed in the 27 Gy Boron dogs irradiated at BMRR. The RBEs and compound factors listed in Table 3 were used to compare the equivalent doses to the skin from the two HFR dogs and two BMRR dogs, see Fig. 2.

Table 3

Experimental groups

Reactor source	Peak phys. dose group (Gy) (range)	Peak eq. dose (Eq*Gy)[a]	Skin phys. dose (Gy)	Skin eq.dose (Eq*Gy)[a]	^{10}B Blood (μg/g)	Number of dogs
BMRR	11	18	9.5	23	0	5
	16.5	27.2	14.3	25.9	0	5
	0	0	0	0	0	2
	15 (12.5-14.6)	8.1-8.6	7.8-8.1	8.8-9.3	31-47	5
	27 (25.5-27.2)	14.8-15.4	15.7-16.4	15.3-16.5	45-60	3
	38 (38.2-38.6)	21.4-24.5	23.3-24.0	22.5-25.8	38-57	3
	64 (64.4-64.5)	34.5-34.5	38.6-38.6	35.9-36.0	62-63	2
HFR	23 (17.5-24.9)	13.0-16.2	9.6-12.3	15.8-17.2	38-62	2

[a]Equivalent dose calculated using RBE for fast neutrons equal to 4.5 (Table 1) and compound factors from Table 2.

DISCUSSION

When BSH was used as boron carrier, boron was essentially excluded from the normal brain parenchyma. The small amount of boron measured in the brain could be largely accounted for by remaining blood volume of the analyzed tissue. The hydrophilic nature of BSH and the tight junctions of the blood brain barrier of the cerebral vasculature would imply that the compound is excluded from the cytoplasm of the capillary endothelium. This would result in a large dose sparing effect for the ^{10}B(n,α)^{7}Li reaction in the capillaries due to its boron distribution giving geometrical effects that include lack of charged particle equilibrium in these small diameter vessels and up and down stream energy depositions not contributing to the endothelial nuclear dose[13]. The boron distribution of BSH in the blood and brain support this hypothesis.

In contrast, the skin contained boron levels similar to that observed in the blood and this would necessitate a large extravascular component for the skin boron values. At this time, the microscopic distribution of the boron is not known in this tissue but this distribution would have a major effect on the dose distribution of the boron capture fission fragments. The compound factor discussed below is an attempt to describe this dose distribution. This factor implies distribution of the boron relative to the critical target cells.

The normal tissue tolerance observed in the EPI dogs revealed important information about the fast neutron components in the epithermal neutron beam. Comparison of the observed skin reactions with those observed for x-rays were difficult and have limitations. The reactions seen in the present study occurred in either none or all of the dogs per dose group. Park[10] had small numbers of dogs and did not have dermal necrosis in 100% of any of his dosage groups. Based on these quantitative differences for the compared bio-logical endpoints, they would imply that in general the calculated RBEs for the fast neutron component in the epithermal neutron beams may be an un-derestimate. The RBE chosen for the nitrogen capture reaction proton can be regarded as high when compared with the range of values reported in the literature[7]. It could be argued that all of the proton reactions including the hydrogen recoil proton have similar RBEs but the literature reveals marked differences in RBE values with relatively minor changes in proton energy[14].

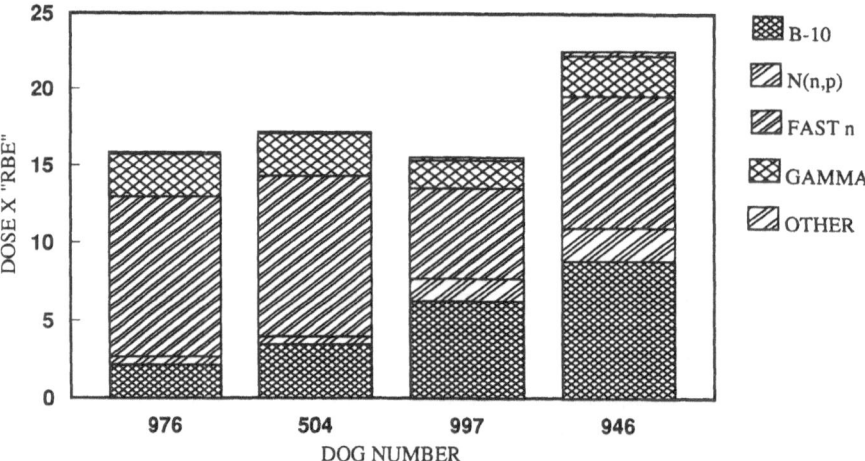

Figure 2. *Skin equivalent dose for 2 dogs irradiated at the HFR (dogs 976 and 504) and at the BMRR (dogs 997 and 946). Dog 997 and 946 were 27 Gy and 38 Gy BORON dogs respectively (see Table 3).*

Therefore, the calculated RBE was limited to the fast neutron induced hydrogen recoil proton and a literature value was used for the the induced nitrogen capture proton. The RBEs obtained for the clinically important late tissue reactions were used for the calculation of equivalent doses.

In this paper, the term fast neutron physical dose is used for all reactions from neutrons with energies greater than 10 keV. There is a lack of information concerning the RBEs of neutrons in the energy ranges of these epithermal neutron beams. However a value of 4.5 or greater is in general agreement with another report[15]. The increase in the apparent RBE with late tissue reactions of the skin compared to the early response in skin is in agreement with the accepted radiation biological premise that early responding tissues have a lower RBE for fast neutrons than late responding tissues[16]. Attempts to date to obtain brain tolerance values for the epithermal neutron beam with acceptable skin reactions have failed, and this has necessitated the use of the same vascular endothelial RBE for the fast neutron dose component in the skin and brain. The peak equivalent dose of the 11 Gy EPI dogs exceeded the reported tolerance for the brain[4]. However, a large fraction of the equivalent dose was due to fast neutrons and their effect diminished rapidly with increasing depth. Therefore, it is proposed that only the surface of the brain was irradiated to the peak dose and that the majority of the brain was exposed to tolerable equivalent doses. Severe atrophy of the irradiated temporal muscle was observed at 12 months post-irradiation. This finding may indicate an increased contribution of the fast neutron component to the human brain since the brain lies much closer to the surface than in the labrador dog. The lower level of the RBE range for late reactions was used for the determination of the compound factor for vascular damage in the brain and subsequent equivalent dose comparisons for individual dog irradiations. The fast neutron RBE and compound factors are obviously indirectly related. Currently many more dogs are being irradiated at both BMRR and HFR to add accuracy and confidence to these estimates prior to any human patient treatment.

The dogs irradiated with boron in the blood showed normal tissue reactions at peak physical radiation dose levels far in excess of those found with conventional x-ray and gamma irradiation. This fact together with the high RBEs for the proton and fast neutron components readily indicate that the compound factor for the boron fission reaction products is less than one. A compound factor equal to 0.5 for skin cannot be confirmed or analyzed further prior to the acquisition of data concerning quantitative microscopic localization data of boron in the skin. The skin concentration is directly related to the blood concentration and the compound factor would be consistent with any blood boron concentration. A compound factor equal to 0.33 for the boron reaction in the cerebral vasculature is in general agreement with mathematical models that predict this degree of sparing due to the inferred boron microdistribution obtained from the boron macrodistribution measured following BSH infusion[17]. The exclusion of BSH from the brain parenchyma independent of blood concentration would indicate the compound factor for the cerebral vasculature would change with the blood concentrations due to microdosimetric considerations[17]. However, assuming a RBE of 2.3, the previously described factor of 0.33 for the effective microscopic capillary dose for the boron capture reaction in the vascular lumen would give a compound factor of 0.76. The compound factor of 0.33 presently observed for the cerebral vasculature suggests an overestimation of the effective microscopic capillary dose for the boron capture reaction by a factor of 2 when compared with previously reported values[18,19].

The fast neutron contamination of the available epithermal neutron beams and the incident and induced gamma components indicate a need for high boron concentrations to effectively use the boron capture reactions for targeted radiotherapy. The fast neutron and gamma dose are non-targeted irradiations and contribute significantly to the normal tissue dose. The fast neutron dose could make the superficial tissues the limiting tissue at low blood boron concentrations. There is significant improvement in the relative dose due to boron capture at depth (tumor location) with increased blood boron concentration, see Fig. 3.

The mixed radiation field results in a complex problem with fractionated treatments. The previously described RBE and compound factor values were used to describe the equivalent dose in skin and at the thermal peak for the HB11 epithermal beam with an 8 cm diameter collimator and a blood boron concentration of 25 ppm. In comparison with a single fraction, an equivalent peak isodose from 4 fractions was calculated assuming 50% repair of the total

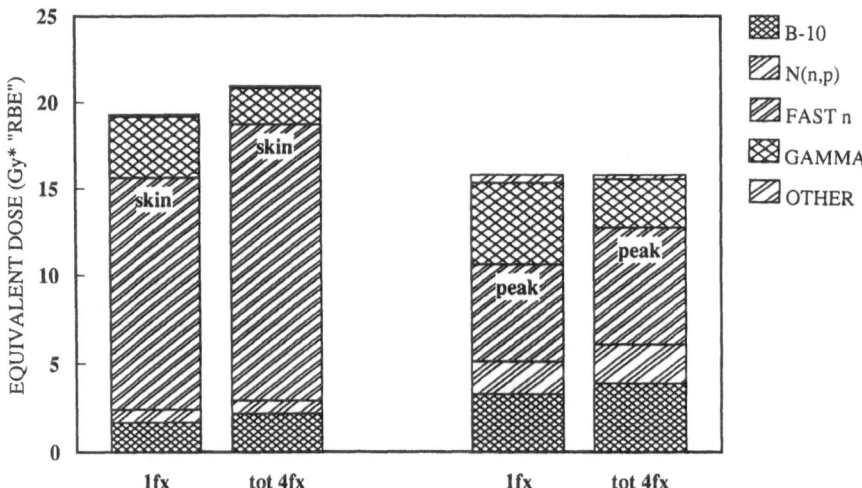

Figure 3. *Change in relative peak physical dose distributions after irradiation at the HFR in relation to boron concentration.*

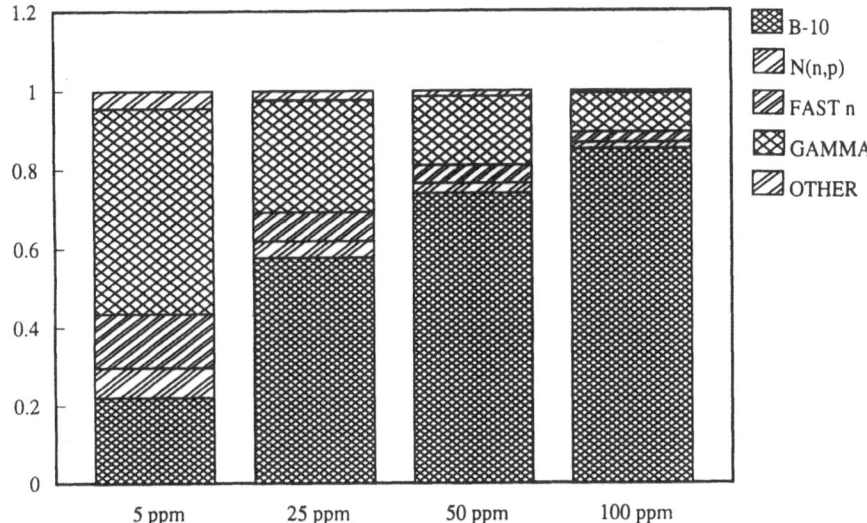

Figure 4. *Skin and peak equivalent dose after irradiation with either a single or 4 fractions at the HFR. Single peak physical dose = 17 Gy. Fractionated total peak physical dose = 20.4 Gy.*

gamma dose and no repair from the relatively low energy fast neutrons. This equivalent isodose at the thermal peak allowed only a 20% increase in the physical dose at peak and was accompanied by a 10% increase in the skin equivalent dose as shown in Fig. 4. If the skin equivalent dose was taken as the isodose, then the fractionated peak equivalent dose would be 10% lower than with a single fraction.

These results stress the complexity of assessing the normal tissue tolerance with boron capture treatments and epithermal neutron beams. Additional large animal model studies using epithermal neutron beams and the various boron carriers are needed to clarify the relative reactions and will give useful definitions of equivalent dose. The radiation oncologist must have this information prior to human patient treatments with this complex bimodal therapy.

ACKNOWLEDGMENT

This study was performed under the auspices of the U.S. Department of Energy, DOE Contract No. DE-AC07-76ID01570.*

---

* The submitted publication has been authored by a subcontractor of the U.S. Government under DOE contract No. DE-AC07-761D01570. Accordingly, the U.S. Government retains a nonexclusive, royalty-free license to publish or reproduce the published form of this contribution, or allows others to do so, for U.S. Government purposes.

REFERENCES

1. Kraft, S.L.; Gavin, P.R.; DeHaan, C.E.; Leathers, C.W.; Bauer, W.F.;
    Dorn III, R.V. The biodistribution of boron in canine spontaneous
    intracranial tumors following borocaptate sodium infusion. In:
    Progress in Neutron Capture Therapy for Cancer. Allen, B.J.; Moore;
    D.E.; Harrington, B.V., eds., Plenum Press, New York, 537-540
    (1992).
2. Nigg, P.W.; Randolph, P.D.; Wheeler, F.J. Demonstration of three
    dimensional deterministic radiation transport theory dose distribu-
    tion analysis for boron neutron capture therapy. *Med. Phys.* 18, 43-53,
    (1991).
3. Fankhauser, H.; Stragliotto, G. Biodistribution of known sulfhydryl
    (BSH) on patients with intracranial tumors. In: *Progress in Neutron
    Capture Therapy for Cancer.* Allen, B.J.; Moore, D.E.; Harrington,
    B.V., eds., Plenum Press, New York, 551-556 (1992).
4. Fike, J.R.; Cann, C.E.; Davis, R.L.; Borcich, J.K.; Phillips, T.L.; Russel,
    L.B. Computed tomography analysis of the canine brain: effects of
    hemibrain x-irradiation. *Radiat. Res.* 99, 294-310 (1984).
5. Zook, B.C.; Bradley, E.W.; Casarett, G.W.; Rodgers, C.C. Pathologic find-
    ings in canine brain irradiated with fractionated fast neutrons or
    photons. *Radiat. Res.* 84, 562-528 (1990).
6. Hopewell, J.W. The skin: it's structure and response to ionizing radia-
    tion. *Int. J. Radiat. Biol.* 57, 751-773 (1990).
7. Archambeau, J.O.; Ines, A. Response of swine skin microvasculature to
    acute single exposures of x-rays: quantification of endothelial
    changes. *Radiat. Res.* 98, 32-51 1984).
8. Gavin, P.R.; Kraft, S.L.; DeHaan, C.E.; Griebenow, M.L.; Moore, M.P. A
    large animal model for boron neutron capture therapy. In: *Progress in
    Neutron Capture Therapy for Cancer.* Allen, B.J.; Moore, D.E.; Har-
    rington, B.V., eds., Plenum Press, New York, 479-484 (1992).
9. DeHaan, C.E.; Gavin, P.R.; Kraft, S.L.; Wheeler, F.J.; Atkinson, C.A.
    Qualitative dose response of the normal canine head to epithermal
    neutron irradiation with and without boron capture. *Radiat. Res. - A
    Twentieth-Century Perspective, Congress Abstracts, Academic Press* 1,
    436 (1990).
10. Park, R.D.; O'Brien, T.R.; Baker, B.B.; Morgan, J.C. Single dose irradia-
    tion of canine skin. *Vet. Radiol.* 15, 108-111 (1974).

11. Gahbauer, R.; Fairchild, R.G.; Goodman, J.H.; Blue, T.E. RBE in normal tissue studies. In: *Boron Neutron Capture Therapy: Toward Clinical Trials of Glioma Treatment*. Gabel, D.; Moss, R.L., eds., Plenum Press, New York, 123 - 128 (1992).

12. Watkins, P. Present status of the three-dimensional treatment planning methodologies for the Petten BNCT facility. In: *Boron Neutron Capture Therapy: Toward Clinical Trials of Glioma Treatment*. Gabel, D.; Moss, R.L., eds., Plenum Press, New York, 101 - 109 (1992).

13. Gavin, P.R.; Wheeler, F.J.; DeHaan, C.E.;, Kraft, S.L.; Moore, M.; Miller, D.L. Dosimetric considerations, radiation tolerance of the normal canine brain following boron neutron capture therapy. *Abstract 38th Annual Meeting Radiation Research Society, New Orleans* (1990).

14. Belli, M.; Cherubini, R.; Finotto, S.; Moschini, G.; Sapora, O.; Simone, G.; Tabocchini, A. RBE-LET relationship for the survival of V79 cells irradiated with low energy protons. *Radiat. Res.* 55, 93-104 (1989).

15. Coderre, J.A.; Makar, M.S.; Micca, P.L., Nawrocky, M.M., Joel, D.D., Slatkin, D.N. Major compound-dependent variations of $^{10}B(n,\alpha)^7Li$ RBE for the rat 9L gliosarcoma *in vitro* and *in vivo*. In: *Int. Workshop on Macro and Microdosimetry and Treatment Planning for Neutron Capture Therapy*. Plenum Press, New York, in the press (1992).

16. Hall, E.T.; Novak, D.K.; Kellner, A.M.; Rossi, H.H.; Morino, S.; Goodman, L.G. RBE as a function of neutron energy. Experimental observations. *Radiat. Res.* 64, 245-255 (1975).

17. Wheeler, F.J. Improved Monte Carlo techniques for neutron capture therapy. In: *Proc. International Workshop on Macro and Microdosimetry and Treatment Planning for Neutron Capture Therap*. Plenum Press, New York, in the press (1992).

18. Rydin, R.A.; Deutsch, O.L.; Muray, B.W. The effect of geometry on capillary wall dose for boron neutron capture therapy. *Phys. Med. Biol.* 21, 134-138 (1976).

19. Kitao,K. Vascular wall dose from boron neutron capture reaction. In: *Boron Neutron Capture Therapy for Tumors*. Hatanaka, H., ed., Nishimura, Niigata, Japan, 191-197 (1986).

PROPOSAL FOR PATIENT SELECTION CRITERIA AND

FOLLOW-UP FOR BNCT IN PATIENTS WITH SUPRATENTORIAL

MALIGNANT GLIOMAS

Heinz Fankhauser and Giuseppe Stragliotto

Department of Neurosurgery
CHUV
CH-1011 Lausanne, Switzerland

INTRODUCTION

Treatment beam and irradiation room of the High Flux reactor in Petten are close to completion and pre-clinical dog irradiation studies were started in summer 1991. It is therefore timely to draft the preliminary protocol which will be applied to the initial patient study using borocaptate sodium (BSH) and epithermal neutrons for the treatment of malignant gliomas. At this stage emphasis is put on inclusion criteria, follow-up, and data collection, rather than on details of BNCT itself. The latter are to be finalized once the results of the irradiation at the treatment position of normal dogs with and without intravenously administered BSH are available.

The main goal of the first patient study is to determine normal tissue tolerance to the epithermal beam at known blood boron levels in patients with maximally resected supratentorial malignant gliomas (SMG), including glioblastoma, anaplastic astrocytoma, and anaplastic oligodendroglioma. Patients undergoing partial resection, or open or stereotactic biopsy may be included into separate groups. Surgical procedures will be combined with intravenous BSH administration and tissue boron analysis. Further information will therefore be gained concerning biodistribution of BSH. We also expect preliminary information concerning the effect of epithermal BNCT on residual or intact, as well as on resected and recurrent tumors.

Inclusion into the study proceeds in two steps, since the decision concerning the BSH uptake study has to be taken before the diagnosis of SMG is histologically confirmed and since the proportion of possible tumor resection cannot always be determined before the operation.

Boron Neutron Capture Therapy, Edited by D. Gable
and R. Moss, Plenum Press, New York, 1992

Group I : Gross total resection, candidate for reoperation

Group II : Partial resection or open biopsy

Group III: Stereotactic biopsy

Figure 1. *Treatment groups according to the surgical procedures in patients with supratentorial malignant gliomas prior to epithermal BNCT with borocaptate sodium (BSH).*

DESCRIPTION OF THE PROTOCOL

After radiological diagnosis of a likely SMG, eligible patients will be tentatively divided into 3 groups (Fig. 1), if they fulfill the basic requirements. In order to be considered for the study, patients have to be scheduled for gross total resection of a lobar tumor who's location is favorable for reoperation in case of post-operative tumor recurrence (group I), for partial resection or open biopsy (group II), or for stereotactic biopsy (group III). They have to be oncologically untreated, except for anticonvulsants and steroids, in good neurological condition (to be defined), aged between 15 and 65, and able to travel to the Netherlands by public transportation. Long term follow-up by the operating physician must be possible. In patients included into group I or group II, consent must be obtained for intravenous administration of BSH prior to the initial surgery. A preoperative MRI must be obtained on all patients.

Open surgery is then performed, coupled with a BSH biodistribution study. The latter is compulsory for groups I and II, but facultative for group III.

If the preoperative diagnosis of likely SMG is histologically confirmed, gross total or partial resection in group I and II patients has to be checked by a CT or MRI enhancement study within 48 hours after surgery. Definite inclusion into the study is now possible if the basic requirements are still fulfilled, including additional requirements (Fig. 2). The latter encompass consent from the patient and his family for BNCT, for regular follow-up, and for reoperation in case of recurrence or radionecrosis in group I (Fig. 3). Consent for autopsy must be obtained from the family only. It must be possible to proceed with radiation within 6 weeks and to make the necessary financial arrangements (Figs. 4 to 6).

For BNCT the patient will travel on public transportation to the Netherlands. The detailed treatment plan will be worked out later, but 4 fractions of BNCT at 24 to 48 hour intervals, after repeat intravenous administration of

---

- **Malignant glioma histologically confirmed**
- **No previous anticancer treatment for the glioma (except steroids, anticonvulsants...)**
- **Early post-operative CT or MRI enhancement study available**
- **Good neurological grade (to be defined)**
- **Age between 15 and 65**
- **Life expectancy of 6 months or more**
- **Able to travel on public transportation**
- **Follow-up possible at original hospital**
- **BNCT possible within 6 weeks after surgery**
- **Patient's consent for BNCT**
- **Family's consent for conditional autopsy**
- **Financial arrangements possible**

Figure 2. *Outline of definite inclusion criteria for all patients.*

BSH are likely to take place. Blood boron levels at the time of radiation of 50 ppm or more will probably be considered optimal for the initial trial.

Clinical and radiological follow-up in the center where the patient has undergone surgery will take place every 6 weeks. The precise work-up, including MRI, neurological, hematological, endocrinological, ophthalmological, ENT, and neuropsychological testing will be defined later, derived from the results of the initial dog radiation experiments and from knowledge of the effects of high and low-LET cranial radiation.

At some stage, the patient is likely to show recurrence or progression on CT or MRI. This may be true recurrence, radionecrosis, or both. At this point, group I patients will be reoperated, again with maximal removal of abnormal tissue. A renewed BSH uptake study will be performed. Besides tumor, vari-

---

- **Gross total resection confirmed by CT or MRI**
- **Tumor location suitable for reoperation**
- **Patient's consent for reoperation in case of recurrence/radionecrosis**
- **Reoperation possible in the same hospital**

Figure 3. *Additional criteria for inclusion of patients into group I : gross total resection and reoperation.*

- Tentative inclusion after likely SMG on CT/MRI
- Gross total resection and BSH distribution study
- Post-operative CT or MRI within 48 hours of surgery
- Definite inclusion following verification of SMG
- Follow-up
- Upon recurrence/radionecrosis : reoperation with
  additional BSH distribution study
- Adjuvant or symptomatic treatment (individual)
- Follow-up
- Autopsy

Figure 4. *Outline of protocol steps for patients in group I: gross total resection and reoperation.*

ous normal tissues will be obtained. In group II and group III, reoperation or re-biopsy is facultative.

Further treatment will depend upon the gross and microscopic findings during reoperation or re-biopsy. This treatment will be left to the decision of the physician in charge. It may include steroids, chemotherapy, repeat surgery, immunotherapy, or others. Further radiation treatment, conventional or interstitial, is strongly advised against.

- Tentative inclusion after likely SMG on CT/MRI
- Partial resection or open biopsy and BSH distribution
  study
- Post-operative CT or MRI within 48 hours of surgery
- Definite inclusion following verification of SMG
- Follow-up
- Upon recurrence/radionecrosis : facultative
  reoperation or biopsy with additional BSH
  distribution study
- Adjuvant or symptomatic treatment (individual)
- Follow-up
- Autopsy

Figure 5. *Outline of protocol steps for patients in group II: partial resection or open biopsy.*

- Tentative inclusion after likely SMG on CT/MRI
- Stereotactic biopsy and facultative BSH distribution
  study
- Post-operative CT or MRI
- Definite inclusion following verification of SMG
- Follow-up
- Upon recurrence/radionecrosis : facultative re-biopsy
- Adjuvant or symptomatic treatment (individual)
- Follow-up
- Autopsy

Figure 6. *Outline of protocol steps for patients in group III: stereotactic biopsy.*

Follow-up after recurrence or radionecrosis will continue until death, with emphasis on function of organs not related to the tumor, such as the eyes, inner ears, pituitary, and on MRI alterations in distant normal brain. The precise methods will be defined later.

Upon death, autopsy is mandatory. Final tissue analysis has to be performed by specially trained pathologists.

Reasons for early modification of this protocol include death from radionecrosis of 2 out of 10 patients within 6 months of BNCT.

DISCUSSION

The primary goal of this study is to determine normal tissue tolerance to epithermal neutrons at various blood levels of ^{10}B. Skin can be observed readily for damage, but the normal tissue most likely to be harmed is the brain. Radionecrosis and tumor recurrence cannot be readily distinguished neither by CT nor MRI. Hence, in order to justify exposure of patients to the potential risks of this new form of radiation and to achieve the objective of the protocol, post-BNCT histology must be obtained. To ensure collection of valid histological information from every single case, this protocol includes reoperation as an integral part for the majority of patients, ie. those in group I. In many neurosurgical centers reoperation is now part of the standard management of malignant gliomas in patients of the younger age group with lobar supratentorial tumors suitably located for repeat maximum resection in case of recurrence. It has been shown to add significant length of survival with acceptable quality of life[2,3]. Systematic reoperation can therefore readily be justified if the patients are properly selected, taking into account tumor location, age, and informed consent. Tumors in eloquent areas or tumors crossing the midline, even if small and occurring in neurologically intact patients, are not to be included into group I.

Radiation within 6 weeks after surgery is required in this first series of patients, anticipating numerous difficulties which may initially be encountered. Once BNCT enters the phase of an efficacy study, shortening of this interval to a maximum of 3 weeks will be required in order to take full advantage of the lowest possible tumor burden.

Autopsy is mandatory, but even if it is an integral part of the protocol, various problems may arise, including death far from the physician in charge, death at an unexpected time, last minute opposition from the family, poor tissue quality due to inappropriate brain and tissue removal or delay between death and fixation. Patient consent for autopsy upon admission to the hospital is routine in some hospitals, but would appear unacceptable for others. The protocol therefore should require initial consent from the patient's family, as condition for inclusion into the study and for BNCT, unless in the further course of the disease the patient himself expresses his opposition. No BNCT should be applied outside this protocol.

Follow-up with detailed examination of all organs and systems at risk for radiation damage is mandatory. Particular attention will be paid to signal alterations on repeat MRI of the brain. Only patients able to attend regular follow-up visits at the original surgical center can be included.

BSH uptake studies at initial operation and reoperation are mandatory for group I and II patients. These will give information concerning the particular patient and the dose of BSH to be administered during actual BNCT, but it will also add to the knowledge of BSH biodistribution in general. The measured tumor concentrations will nevertheless not be used as criterion for definite inclusion into the study, at least as long as BNCT is limited to malignant gliomas. This may change if lower grade tumors are considered for BNCT.

Besides group I, two further groups are defined for patients with tumors unsuitable for gross total resection and reoperation: group II for those undergoing partial open surgery and group III for those undergoing stereotactic biopsy. It is important to accept that radical operation in group II and III patients is withheld because of the unfavorable location of their tumor, e.g. deep seated gliomas or gliomas crossing the midline. If radical surgery is contraindicated for other reasons, such as large size of tumor, old age, bad neurological or general condition, these patients cannot be included into the protocol.

A BSH uptake study is recommended even in patients undergoing stereotactic biopsy, but this is not compulsory. Stereotactic biopsy yields a few very small tissue fragments; above all they must be used to guarantee a reliable histological diagnosis. Little or no material will be left for boron assessment, which will be difficult and less reliable. Nevertheless, valuable pharmacokinetics information can still be gained from blood sampling in patients undergoing a BSH study during stereotactic biopsy.

Definite inclusion into the study and into one of the three groups is possible only after the primary surgical procedure, since histological confirmation of the diagnosis of SMG is necessary. As this is not a randomized study, it is still possible at this time to move patients from group I to group II or vice versa, according to the findings on early postoperative CT or MRI. Neuroradiological examination without and with intravenous contrast medium has to be performed in groups I and II within 2 days after the operation. This will verify if the pre-operatively enhancing tissue has been resected, i.e. gross total resection, or if substantial amounts of enhancing tissue have been left behind. It is known that during the first 3 days, the surgical trauma does not induce any contrast enhancement. All enhancement seen within this period therefore represents preoperatively enhancing tissue[1]. If CT or MRI examinations are done later, visible contrast enhancement may represent tumor or surgically induced alterations, including neovascularity.

This protocol is designed to provide essential information concerning normal tissue tolerance to epithermal BNCT in humans as rapidly as possible, with a maximum of non-ambiguous data from various tissues. The inclusion of reoperation and autopsy as an integral part of the study may be considered excessive. Reoperation in selected cases is part of standard management even at present time[2,3] and therefore does not impose an extra-burden upon the patients. Together, reoperation and autopsy offer a unique chance to study various tissues at different times after BNCT. In this manner, every individual patient provides most valuable information on its own, and no larger series are necessary for identification of adverse treatment effects by statistical methods. Additional inclusion of patients with residual or intact tumor after the initial surgical procedure will allow to draw preliminary conclusions concerning the efficacy of BNCT against gliomas, but this is not the main goal of the study. Therefore, absence of detectable action against tumor recurrence at this stage should by no means be interpreted as proof of inefficacy of BNCT. This question can only be systematically approached after the tolerable levels of BNCT by normal tissues have been worked out.

ACKNOWLEDGEMENTS

The authors would like to thank Prof. Patrick R. Gavin, Washington State University, Pullman, WA. for his critical review of the manuscript. This work has been supported by the Fonds National de la Recherche Scientifique, the Ligue Suisse contre le Cancer, the Ligue Vaudoise contre le Cancer, and the Commission of the European Communities.

REFERENCES

1. Jeffries, B.F.; Kishore, P.R.S.; Singh, K.S.; Ghatak, N.R.; Krempa, J. Contrast enhancement in the postoperative brain. *Radiology* 139, 409-413 (1981).

2.   Ammirati, M.; Galicich, J.H.; Arbit, E.; Liao, Y. Reoperation in the
     treatment of recurrent intracranial malignant gliomas. *Neurosurgery*
     21, 607-614 (1987).
3.   Harsh, G.R.; Levin, V.A.; Gutin, P.H.; Seager, M.; Silver, P.; Wilson,
     C.B. Reoperation for recurrent glioblastoma and anaplastic astrocyt-
     oma. *Neurosurgery* 21, 615-621 (1987).

A PROPOSAL FOR PHASE I CLINICAL TRIALS OF

GLIOMA PATIENTS

Harry Bartelink

Department of Radiotherapy
The Netherlands Cancer Institute
NL-1066 CX Amsterdam, The Netherlands

INTRODUCTION

With the renewed interest in Boron Neutron Capture Therapy (BNCT) for glioma patients, the European Collaboration on BNCT will soon embark on a clinical pilot study in patients with high grade glioma. This pilot study will be twofold, consisting of a pharmacokinetic study of $Na_2B_{12}H_{11}SH$ (BSH) and of a radiation dose escalating study.

STUDY DESIGN AND DISCUSSION

Pharmacokinetic Study

Firstly, a time line study will be performed. [10]B-enriched BSH will be administered in a dose of 25 mg [10]B/kg during one hour infusion to subgroups of 8-10 patients at 3, 6, 12 or 18 hours prior to intraoperative tumour biopsy for histology and [10]B analysis. Blood samples will be obtained at short time intervals and 24 hour urine samples will be collected for boron determination. The techniques available to use are prompt gamma-ray spectroscopy (PGRS), inductively coupled plasma-atomic emission spectroscopy (ICP-AES) and quantitative neutron capture radiography (QNCR). The results of this pilot study will give the clearance pattern of the boron compound, e.g. whether it is model dependent, i.e. 2- or 3-compartment model[4] or biphasic, as has been observed in patients[1,2,3]. It will also allow us to choose the most optimal time point of BSH administration, that is at the most favourable tumour-to-blood ratio of [10]B concentrations.

Subsequently, increasing doses of BSH will be administered (e.g. 25, 50, 75 mg ^{10}B/kg) at the "optimal" time interval. Again, blood and urine samples will be collected for boron analysis. This should allow us to determine the relationship between administered dose of BSH and the boron concentrations in the serum and in the tumour.

A Radiation Dose Escalating Study

The aim of a radiation dose escalating study is to determine the "tolerance" dose of the human brain to BNCT. Of prime importance for such a study is the availability of a reliable, non-invasive and non lethal endpoint for assessing brain damage in the patient to irradiation. At present, ^{157}Gd-DTPA MRI as an assay for blood brain barrier (BBB) break down appears to be the most promising test[4]. Prior to clinical application, a reliable dose response relationship will have to be obtained in dogs with the epithermal neutron beam at the high flux reactor in Petten (HB11). Also the effect of fractionation on the brain tolerance will have to be assessed in these animals. Once sufficient information is obtained from these experimental studies, a radiation dose escalating study will be started in glioma patients after radical removal of the primary. Presumably the epithermal neutron irradiations will be given in 5 to 6 fractions at 24-48 hour intervals with BSH administered at a given time before each fraction.

Ideally, the BSH dosage should be adjusted based on the serum level of boron, as determined by PGRS, or, alternatively, the time point of irradiation will be shifted, in order to irradiate the patient each time at the same serum ^{10}B concentration. Increasing total radiation doses will be delivered using ^{157}Gd-DTPA MRI as an endpoint of non-lethal radiation damage to the brain. It will be important to know from the dog experiments whether BBB break-down occurs within 2 weeks soon after epithermal neutron irradiation in the presence of BSH. If so, leakage of BSH into the perivascular space may lead to significant radiation exposure to other cells than the endothelial cells and, consequently, may change significantly the level of brain damage to neutron irradiation as well as the predominant underlying pathophysiological process. The success of BNCT will be largely determined by obtaining good insights in these underlying processes of radiation damage to the brain.

In this radiation dose tolerance study, the therapeutic ratio might be further improved by delivering the irradiation through 3 or 4 beam portals rather than by simple uni- or bilateral fields. This requires an extensive quality control study, in which a computer treatment planning system designed for BNCT will be used for phantom and in vivo dosimetry. It is believed that with the joint efforts between physicists, radiobiologists and radiation oncologists and with a good interactive collaboration, such a phase I clinical trial should lead to a well designed and scientifically sound phase III trial in glioma patients.

ACKNOWLEDGEMENTS

The European Collaboration on Boron Neutron Capture Therapy is supported by the Commission of European Communities.

REFERENCES

1. Haritz, D.; Piscol, K.; Gabel, D. The distribution of BSH in patients with malignant glioma. In: *Progress in Neutron Capture Therapy for Cancer*. Allen, B.J.; Moore, D.E.; Harrington, B.V., eds., Plenum Press, New York, 557-560 (1992).
2. Stragliotto, G.; Munafo, A.; Biollaz J.; Fankhauser, H. Pharmacokinetics of boron sulphydryl (BSH) in patients with intracranial tumors. In: *Progress in Neutron Capture Therapy for Cancer*. Allen, B.J.; Moore, D.E.; Harrington, B.V., eds., Plenum Press, New York, 549-550 (1992).
3. Sweet, W. Supplementary pharmacological studies between 1972 and 1977 on purified mercaptoundecahydrododecaborate. In: *Boron Neutron Capture Therapy for Tumors*. Hatanaka, H., ed., Nishimura, Niigata, Japan, 59-76 (1986).
4. Gavin, P.R.; Kraft, S.L.; DeHaan; C.E., Griebenow, M.L.; Moore, M.P. A large animal model for boron neutron capture therapy. In: *Progress in Neutron Capture Therapy for Cancer*. Allen, B.J.; Moore, D.E.; Harrington, B.V., eds., Plenum Press, New York, 479-484 (1992).

APPROACHES TO THE DESIGN AND EVALUATION

OF COMPOUNDS FOR BNCT

Albert H. Soloway, Rolf F. Barth, Liang Liu,
Werner Tjarks, Iwona M. Wyzlic,
and Abdul K.M. Anisuzzaman

Colleges of Pharmacy and Medicine
The Ohio State University
Columbus, Ohio 43210, USA

INTRODUCTION

The development of new and more effective agents for neutron capture therapy (NCT) continues to remain an important and ongoing problem as we strive to prepare compounds which possess greater and greater specificity for tumor versus contiguous normal tissues[1]. The problem that the synthetic chemist encounters in the rational development of such compounds is the paucity of information available as to the biochemical and physiological differences between normal and malignant cells that will allow the selective incorporation of the neutron absorber into tumor cells and their stroma. Cancer chemotherapeutic agents[2] which have been used clinically for more than 40 years, have been evaluated for their tumoricidal activity and not on the basis of concentration differentials that may exist between tumor and adjacent normal tissues. On the other hand, concentration differentials have been very important in the development of radiopharmaceuticals, which have been used largely as diagnostic agents for malignancies, but in such instances only trace amounts, from a chemical standpoint, are administered[3]. Therefore, it is understandable that their clinical pharmacokinetic parameters may be considerably different from those observed, where significant amounts of compound are administered. This is certainly the requirement for NCT. Thus, the key questions for the development of new compounds for NCT are: (1) what is the rational basis for designing specific tumor-targeting agents, and (2) how should such structures be evaluated from an *in vitro* and *in vivo* standpoint in order to ascertain their potential utility in NCT?

Early studies in the evaluation of chemical compounds for NCT were largely empirical. Since the initial focus of NCT was for the treatment of

Boron Neutron Capture Therapy, Edited by D. Gable
and R. Moss, Plenum Press, New York, 1992

malignant brain tumors, compounds excluded from normal brain by the blood-brain barrier initially appeared to be the most attractive. However, it became apparent that such a criterion was insufficient since it was also necessary that there be suitable concentration differentials between tumor and blood. It was on this basis that $Na_2B_{12}H_{11}SH$ was identified from studies in brain tumor-bearing animals[4], and is being used clinically in Japan[5]. For NCT to contribute to the treatment of glioblastoma multiforme, it is now clear that tumor fibrils and cells that have infiltrated normal brain and indeed may be protected by the blood-brain barrier must be targeted for destruction. Otherwise, these sites, unobservable to the clinicians, will serve as the foci for tumor regrowth.

The developing area of photodynamic therapy (PDT) is one that is closely related to NCT in its chemical requirements and therefore can provide some direction for the synthesis of new neutron absorbers[6]. This has been the basis for the preparation of boron-containing porphyrins and phthalocyanines[7,8]. However, PDT, as well as the development of classical radiation sensitizers[9], remain in an early stage in the preparation of new and more effective tumor-specific agents and in understanding the biochemical/physiological basis by which existing structures achieve specificity. This has been a major limitation in the development of new tumor-selective compounds, i.e. the lack of involvement of molecular pharmacologists and biochemists in understanding the mechanism by which current agents achieve localization.

Efforts have been underway at a number of laboratories to develop new and more specific NCT agents. These have included the synthesis of boron-containing compounds which can be divided into three broad categories, as listed in Table 1.

A number of the compounds listed in Table 1 have now been synthesized and are in varying stages of development. An important question is how are these compounds to be evaluated in order to select those agents with good potential for use in NCT? A major limitation for the synthetic chemist is that new compounds, either of low or high molecular weight, are initially prepared in milligram quantities. It would be very wasteful and time-consuming to prepare gram amounts of compound, for example from a five or more step synthetic sequence only to determine that the compound possessed little utility for NCT. The development of a screening process which would use milligram amounts of material would offer clear advantages. Only if the compound demonstrated promise, would sufficient amounts be synthesized for evaluation in tumor-bearing animals. With this objective in mind, we have undertaken to screen compounds initially by an *in vitro* tumor cell preparation.

Establishing an evaluative procedure for a compound through a sequence of steps is an appropriate and necessary process to permit the comparison of various chemical entities that are to be considered for use in BNCT. Such an *in vitro* procedure will not only permit the evaluation of a particular compound but will allow a correlation of chemical structure with the extent of cellular uptake and persistence.

Table 1

Boron-Containing Structures for BNCT

1. Cellular Building Blocks

a. Amino Acids/Protein Precursors
b. Purines, Pyrimidines and Intermediates
    in Nucleic Acid Synthesis
c. Precursors of Membrane Lipids

2. Tumor Cell Incorporators/Binders

a. Porphyrins/Phthalocyanines
b. Monoclonal Antibodies
c. Lipoproteins and liposomes

3. Other Compounds

a. Anticancer Agents (i.e. alkylating agents)
b. Sulfhydryl Derivatives/Thiouracils
c. Radiation Sensitizers
d. CNS Depressants (Phenothiazines, Hydantoins)
e. Receptor Binders and DNA Intercalators
f. Oligonucleotides - Antisense Agents

These *in vitro* studies are not meant to be a substitute for *in vivo* studies. They can provide information as to whether a compound under these more idealized conditions can become incorporated and retained within tumor cells. Comparable studies with normal cells would provide some indication as to whether there was selectivity for the malignant cells. However, there are clearly major limitations to a total reliance upon the *in vitro* preparation as the sole screening entity. Metabolism, elimination via the reticuloendo-thelial system, conjugation to macromolecular structures and transport bar-riers are but a few of the factors that are of much less importance in cell preparations than in mammalian systems. This is readily apparent when one compares the cellular uptake of boronated antibodies *in vitro* to their *in vivo* distribution in tumor-bearing animals[10]. This difference does not totally negate the information which the cell preparation can provide, especially for low molecular weight compounds that may not undergo rapid enzymatic transformations. For such structures, the degree and extent of protein binding may be of significance since the cellular transport mechanisms of such con-jugates may differ considerably from their unconjugated counterparts. The ability to analyze aliquot of cells in an amount as small as 10^7 cells by atomic emission spectroscopy has permitted both tumor cell uptake and persistence studies of various boron compounds. Preparations containing amounts of boron of 1 microgram or greater in such cell samples are both measurable and significant. Tissue preparations are solubilized by concentrated sulfuric acid and cell structure disrupted by hydrogen peroxide. After appropriate dilution, aliquots are injected directly into the spectrometer[11]. One advantage of this analytical technique is the fact that there is no need to prepare the compound as its B-10 enriched derivative since the measurement is for total boron by ICP or DCP atomic emission spectroscopy. The synthesis of B-10 enriched compounds would occur at the time of their planned use in radiation studies. This research would occur subsequent to demonstrations that cellular uptake occurs and persistence is observed under *in vitro* conditions. Cellular persistence for a particular compound would warrant the subcellular fractionation of various organelles to determine the cellular compartment where the boron compound has localized. This could be determined by differential centrifugation after rupture of cell membrane. Alternatively, it is now becoming practical by Electron Energy Loss Spectroscopy (EELS)/Electron Spectroscopic Imaging (ESI)[12] and Secondary Ion Mass Spectrometry (SIMS)[13] to determine the subcellular localization of various boron delivery agents and the subcellular differentials that exist. This information may be of use in explaining the compound's cellular toxicity and in estimating the potential effects in radiobiological studies. Knowing precisely the cellular distribution pattern, it should be possible to correlate this information with the observed radiation effects. In this manner, a practical method for determining RBE for BNCT in cell cultures may become possible. Also, the effect of BNCT in cell culture may be a useful parameter before undertaking *in vivo* animal radiation studies.

Compounds which are to be evaluated in tumor-bearing animals will require the assessment of the compound's toxicity in the same species in which

Table 2

In vitro *Cell Culture Studies*
Evaluation of Compound with
Semiconfluence Cell Culture Preparation

1. Measurement of cellular boron uptake
2. Persistence studies
3. Determination of subcellular fraction content (ultracentrifugation, EELS, SIMS)
4. Assessment of cellular toxicity
5. Radiobiological studies

the tumor is to be transplanted. Significant toxicity could affect the pharmacokinetic pattern. Thus, such studies may serve to establish what are the nontoxic doses which will be administered systemically to tumor-bearing animals.

As an intermediate step between *in vitro* cell culture and systemic administration of the compound into tumor-bearing animals, compounds can be administered by intratumor or perilesional injection. The major purpose of this mode of administration will be to assess a compound's tumor persistence under *in vivo* conditions.

Table 3

In vivo *Studies in Tumor-Bearing Animals*

1. Compound evaluation by intratumor or perilesional tumor administration - uptake and persistence studies
2. Intraperitoneal, intravenous and intraarterial administration - pharmacokinetic studies
3. BNCT of animal tumors and comparison with other irradiation procedures

It will, of course, be important to eliminate any artefacts that could arise, such as the precipitation and insolubility of the compound in tissue. This could give rise to concentrations of the compound at the site of administration and be unrelated to selective tumor cell incorporation. If the compound demonstrates cellular persistence both under *in vitro* and perilesional administration, it should be evaluated systemically. An important question is the appropriate mode of administration and whether constant blood levels are important (i.e. sustained release preparations) in compound evaluation? Pharmacokinetic studies in tumor-bearing animals are a necessary prelude before contemplating any radiation procedures involving BNCT in tumor-bearing animals.

At this juncture in the evaluation procedure and with animal irradiation studies planned, it will be necessary to have available the compound in its ^{10}B enriched form. From the compound's subcellular location and concentration, the radiological contribution of various components in the impinging neutron beam and calculations as to the length of time needed for the irradiation procedure, it will be possible to determine the expected radiation to the tumor and surrounding structures. BNCT irradiation in tumor-bearing animals, especially smaller animals, suffers from the difficulty to confine thermal and epithermal neutrons specifically to the tumor area even using specially fabricated shielding devices. Unfortunately, whole body irradiation can clearly occur even if limited to a certain extent. This could compromise the effectiveness of BNCT in animal tumor treatment, especially if tissues such as the gastrointestinal tract or liver possessed high boron concentrations. As a consequence, radiotoxicity would be observed in such instances. However, under clinical conditions either such tissues would not be exposed to significant neutron fluences or these sensitive structures may be adequately shielded. This problem of developing a suitable model for evaluating BNCT in tumor-bearing animals has been a continuing one and stems directly from an inability to focus the neutron beams to the desired tissue even with the use of more energetic epithermal neutrons.

This limitation in the use of animal models has led to the crucial need for involving radiation therapists in clinical treatment planning. The clinical studies which they would suggest would be a logical follow on from the animal research described above. It must include initially clinical distribution studies of a particular compound, using established pharmacokinetic principles, of safe dosages to patients undergoing tumor resection. From such information, it will be possible to determine that a sufficient radiation dosage is delivered to the tumor in order to achieve a therapeutic effect, comprising all radiation components, while ensuring that normal tissues in the beam path would be spared unacceptable radiation effects. This requirement is a *sine qua non* before any clinical BNCT studies can be initiated.

Table 4

Clinical BNCT Trials

1. Macro and microdosimetry
2. Synthesis of B-10 enriched compounds
3. Treatment planning factors
4. Clinical protocol
5. Clinical BNCT studies

It is understandable that radiotoxicity to normal tissues is totally inappropriate and was the basis for the early failure of BNCT in the United States. Of major importance as well, is that there be a measured and cytotoxic dose to the tumor. Thus, there must be both safety and efficacy. To do otherwise would perpetuate among radiation therapists and other clinicians, a lack of confidence in the potential utility of BNCT in managing patients with cancers that are untreatable by existing modalities.

In conclusion, this presentation has sought to develop a logical basis for the design of new boron compounds for BNCT that is supported by biochemical and physiological differences existing between normal and malignant tissues. Of great importance as well, is the methodology by which these new entities are to be evaluated both by *in vitro* and *in vivo* procedures. This information which has been outlined will serve to provide the biological basis and justification for the ultimate use of new and more tumor selective agents in clinical BNCT trials.

ACKNOWLEDGEMENTS

Our research has been supported by the Department of Energy grant DE-FG02-90ER60972 and contract DE-AC02-76CH000616, National Cancer Institute grant 1RO1 CA53896 and a grant from The Ohio State University Comprehensive Cancer Center.

REFERENCES

1. Barth, R.F.; Soloway, A.H.; Fairchild, R.G. Boron neutron capture therapy of cancer. *Cancer Res.* 50, 1061-1070 (1990).
2. Pratt, W.B.; Ruddon, R.W. The Anticancer Drugs. Oxford Univ. Press, New York, Oxford (1979).
3. R.P. Spencer. Radiopharmaceuticals - Structure activity Relationships. *Grune and Stratton, Inc., New York* (1981).
4. Soloway, A.H.; Hatanaka, H.; Davis, M.A. Penetration of brain and brain tumor. VII. Tumor-binding sulfhydryl boron compounds. *J. Med. Chem.* 10, 714-717 (1967).
5. Hatanaka, H. Clinical results of boron neutron capture therapy. In: *Neutron Beam Design, Development, and Performance for Neutron Capture Therapy.* Harling, O.K.; Bernard, J.A.; Zamenhof, R.G., eds., Plenum Press, New York, 15-20 (1990).
6. Dougherty, T.J. Photodynamic Therapy - New Approaches. *Sem. Surg. Oncol.* 5, 6- (1989).
7. Kahl, S.B.; Joel, D.D.; Nawrocky, M.M.; Micca, P.L.; Tran, K.P.; Finkel, G.C.; Slatkin, D.N. Uptake of a *nido*-carboranylporphyrin by human glioma xenografts in athymic nude mice and by syngeneic ovarian carcinomas in immunocompetent mice. *Proc. Natl. Acad. Sci. USA* 87, 7265-7269 (1990).
8. Miura, M.; Gabel, D.; Fairchild, R.G.; Laster, B.H.; Warkentien, L.S. Synthesis and *in vivo* studies of a carboranyl porphyrin. *Strahlenther. Onkol.* 165, 131-134 (1989).
9. Brown, J.M. Keynote address: hypoxic cell radiosensitizers: where next? *Int. J. Radiat. Oncol. Biol. Phys.* 16, 987- (1989).
10. Barth, R.F.; Soloway, A.H.; Adams, D.M.; Alam, F. Delivery of boron-10 for neutron capture therapy by means of monoclonal antibodies: an Assessment of current realities and future prospects. In: *Progress in*

Neutron Capture Therapy for Cancer. Allen, B.J.; Moore, D.E.; Harrington, B.V., eds., Plenum Press, New York, 265-268 (1992).

11. Barth, R.F.; Adams, D.M.; Soloway, A.H.; Mechetner, E.; Alam, F.; Anisuzzaman, A.K.M. Determination of boron tissues and cells using direct-current plasma atomic emission spectroscopy. *Anal. Chem.* 63, 890-893 (1991).

12. Bendayan, M.; Barth, R.F.; Gingras, D.; Londono, I.; Robinson, P.T.; Alam, F.; Adams, D.M.; Mattiazzi, L. Electron spectroscopic imaging for high-resolution immunocytochemistry: use of boronated protein A. *J. Histochem. Cytochem.* 37, 573-580 (1989).

13. Ausserer, W.A.; Ling, Y.C.; Chandra, S.; Morrison, G.H. Quantitative imaging of boron, calcium, magnesium, potassium, and sodium distributions in cultured cells with ion microscopy. *Anal. Chem.* 61, 2690-2695 (1989).

THE IMPLEMENTATION STRATEGY FOR

BNCT TRIALS IN AUSTRALIA

Barry J. Allen

Biomedicine and Health Program
Australian Nuclear Science and Technology Organisation
Menai NSW 2234, Australia

INTRODUCTION

The Australian research program in Neutron Capture Therapy (NCT) had its beginnings in 1982[1,2] following a visit by the author to Professor Hatanaka in Japan. While the commitment by Ansto has been restrained and relatively constant in recent years at 2-3 person years pa, that of university and medical collaborators continues to grow. This growth in Australian interest has been mirrored by that in the USA, Europe and Japan, and symposia and workshops are now common on the international scene.

Our research and development program has utilised available facilities in Australia and has relied on international collaborations to complement our expertise. We have contributed to symposia and workshops and over 40 papers have been published.

While NCT in itself is a most stimulating and rewarding research subject, the ultimate objective is to bring a new therapeutic modality into clinical practice. With this in mind, the development of a patient epithermal NCT facility at the 10 MW High Flux Australian Reactor (HIFAR) is our ultimate objective. If this project fails, then clinical NCT will not be available in Australia until the development of intense accelerator neutron sources.

This paper reviews the multifaceted research effort in Australia, and the implementation strategy for BNCT trials at the proposed HIFAR epithermal NCT facility.

Boron Neutron Capture Therapy, Edited by D. Gable
and R. Moss, Plenum Press, New York, 1992

SYNTHESIS OF BORON COMPOUNDS

Our initial interest lay in the synthesis of melanoma affined boron compounds. Boron containing monoclonal antibodies against melanoma and leukaemia were synthesised at Sydney University and found to contain up to 1500 boron atoms per antibody without significant loss of specificity[3].

Decarborane derivatives of thiouracil were synthesised at Ansto, some with very poor solubility[4,5]. However, selective uptake was observed when a negative charged liposome carrier was used[6].

Current research lies in the synthesis of a carborane phenylalanine at Sydney University; gadolinium and boron derivatives of DNA ligands at Melbourne University and the Peter MacCallum Cancer Institute[7]; and Pt and Ru based compounds at the Australian National University.

CELL LINES

The murine melanoma cell lines B16 and Harding Passey (HP) were obtained from Japan and USA, and a number of human melanoma lines from several Australian research institutes. In vitro neutron capture experiments were undertaken with three human lines and later the B16 line and survival curves after incubation in boronophenylalanine were obtained[8,9,10].

More recently melanoma cell lines with selective metastatic potential and a pancreatic cancer line have been obtained from the USA.

IN VIVO MODELS

From the outset, the nude mouse model was adopted to provide a test bed for human melanomas at Ansto. Most BNCT studies relate to murine melanoma in syngeneic mice and might not reflect the varied radiation response which is typical of human melanoma. We were able to report the first long term regression of Harding Passey (HP) melanoma in nude mice after BNCT in the Moata reactor[11]. Mice survived up to 12 months post BNCT before termination. The HP melanoma was obtained from BNL and is an excellent model, being very stable and reliable. A study of BPA uptake in HP tumours showed no dependence of tumour boron concentration on tumour size up to 350 mg.

While the uptake of BPA by HP tumours may not be typical of the human melanoma lines in our possession, it is widely used overseas and provides a good gauge of the selectivity of melanoma affined boron compounds. The BALB/c mouse is syngeneic for HP melanoma, and this model is also being studied at Sydney University.

Our biodistribution studies have centered on BPA·HCl and more recently BPA·fructose administered by ip injection. The strong peak in tumour boron at 4 hours in the tumour/normal tissue ratio observed with BPA·HCL in the HP nude mouse model is not observed with BPA·fructose.

Other studies relate to the biodistributions of boron containing thiouracils in collaboration with the University of Bremen[12], monoclonal antibodies against melanoma and leukaemia with Royal Newcastle and Westmead hospitals[13] and liposomes[6].

In addition, an intrahepatic and intrapancreatic model in nude mice has been established in collaboration with Pavia University with the HP line and preliminary biodistributions obtained[14]. Results show that the intrapancreatic tumour growth rate exceeds that of the intrahepatic and subcutaneous lesions. BPA·fructose uptake in the internal xenografts was well in excess of that of the liver but less than that of the pancreas.

In collaboration with the University of California San Francisco, enhanced uptake of a boron porphyrin derivative (BOPP) and its cellular incorporation in isolated cells has been demonstrated in a rat glioma model at Royal Melbourne Hospital[15]. Confirmation of boron uptake remains to be done.

A rat mammary carcinoma model at Royal North Shore Hospital is being tested for uptake by various boron compounds. Initial studies with BPA·fructose show no enhanced incorporation in the tumours and we are now investigating the pharmacokinetics of BOPP in this model.

Spontaneous tumours in dogs are now included in our biodistribution studies. Boron compounds are administered before surgery and tumour samples analysed for boron. Cancers include oral melanoma, osteosarcoma and fibrosarcoma.

While animal models are useful in selecting the most effective boron compounds, the human patient is the ultimate test of efficacy. In Australia, hospital ethics committee approval is required for procedures based on individual patient use (IPU). Biodistribution studies are planned with BPA·fructose in patients with high grade brain tumours (Prince Henry Hospital and Prince of Wales Children's Hospital) and recurrent or metastatic melanoma (Royal Prince Alfred Hospital). The first batch of compound has recently been tested for pyrogens and pathogens, as administration will be by iv infusion of 1 g BPA·fructose at 1-12 hours before surgery. The first patient studies are scheduled for September, and will complement the BNL data obtained with oral administration of BPA.

ANALYTICAL TECHNIQUES

Inductively coupled plasma atomic emission spectroscopy (ICP-AES) for boron analysis was developed at Ansto[16] and is used in all biodistribution studies. Current improvements relate to microwave tissue digestion. Sensitivity is 15 ppb.

Boron Neutron Capture Radiography is available and is used to study the spatial distribution (10-20 μm) of boron in tumour and tissue samples. In a collaborative study with Harwell and Oxford, NCR of a human melanoma section was found to be superior to results obtained with conventional immunohistological techniques.

To investigate subcellular boron distributions, boron electron energy loss spectroscopy in transmission electron microscopy is being developed at Sydney University[17,18]. Electron microscopy studies of changes of melanoma cell morphology following boron neutron capture were not able to distinguish between damage caused by neutrons alone[19,20]. However, high LET radiation does induce morphological changes in surviving cells.

The in vivo pharmacokinetics of BSH is being studied at the University of NSW by Boron Magnetic Resonance Spectroscopy[21.] A small coil is placed over the sc tumour in an anaesthetised rat, located in a tube at the centre of a 4.7 T magnet. The time course of uptake and leakage of the boron compound from the tumour can then be studied. Unfortunately BPA and BTPP provide broadened signals in vitro and are not suitable for in vivo studies.

The morphology and vascularisation of tumours of different cell lines are being examined and related to the pharmacokinetic properties and recurrence rates after treatment. Of primary concern is the effect of NCT on the microvasculature. While the geometric factor of three should apply for capillary dose reduction from blood borne boron, experimental data are not yet available which confirm this effect. We plan to investigate these effects by injection replication techniques in collaboration with the Peter MacCallum Cancer Institute[22].

IN VITRO AND IN VIVO IRRADIATION FACILITIES

The 100 kW Argonaut reactor Moata is used for in vitro and small animal in vivo NCT experiments[23]. A 20 cm Bi filter is located in the centre of the thermal column. Neutron shielding behind the Bi defines a void in which 12 vials can be rotated in a vertical plane in a neutron field uniform to 5%. Three nude mice in ^{6}LiF epoxy shield tubes in an isolation capsule can be placed behind the Bi in a flux of 10^{10} cm^{-2} s^{-1}. The incident gamma dose is 30% of the total absorbed dose. Therapeutic neutron capture irradiations can be obtained in 20 minutes for 20 ppm boron in tumours.

This facility was used to demonstrate the regression of HP melanoma xenografts in the nude mouse model after NCT with ^{10}BPA ip administration. While a fluence of 10^{13} cm^{-2} caused a 3 week growth delay in the melanomas, no local cures were obtained. However, after ip administration of ^{10}BPA·HCl four hours before neutron irradiation, 6 of 8 tumours showed long term regression, one tumour recurring at 280 days post BNCT. Some mice survived for up to a year before sacrifice. In most cases a black tumour bed remained under the skin.

Histological examination of tumour bed sections with haematoxylin & eosin, S100 monoclonal antibody, and avidin-biotin immunoperoxidase stains identified, in some cases, that single melanoma cells were present in the midst of macrophages and tumour debris[24]. These cells had not divided for over 300 days and have either lost the capability for mitosis or are in an hypoxic state, caused by the surrounding tumour debris.

BPA is indicated for palliative NCT of melanoma metastatic to the brain. However, the dopamine and noradrenaline tracts might be expected to exhibit enhanced incorporation of BPA and could be preferentially damaged. Boron concentrations in the mid-brain, cortex and cerebellum of mice are found to be comparable to those in muscle, indicating that while BPA passes through the blood brain barrier, there is no average preferential uptake. While this result is encouraging, NCR is needed to determine whether the dopamine and noradrenaline tracts exhibit enhanced uptake of BPA at the microscopic level.

The effect of NCT after ip administration of ^{10}BPA·HCl on the blood brain barrier has been investigated in mouse brains. Boron concentrations after BPA administration before and after NCT showed a significant increase in the cerebellum only.

The Moata NCT facility was also used to study the effect of DNA bound ^{157}Gd. After neutron capture, ^{157}Gd emits very short range Auger electrons which can induce double strand breaks in circular DNA[25,26]. The possible role of Gd for in vivo NCT has also been considered[27].

NEUTRON AND GAMMA TRANSPORT CALCULATIONS

Calculation of the effects of various filters and filtered beams on neutron dose-depth distributions[28,29,30,31,32] show that an epithermal beam holds the promise of out-patient treatment by NCT which would lead to much improved prognosis and clinical acceptability. Various codes employed at Ansto include DOT, DORT, and MCNP. The codes are now available for use in the UNIX operating system on the Fujitsu VP2200 supercomputer at Lucas Heights.

HIFAR EXPERIMENTS

The thermal flux, Cd ratio and gamma dose rate have been measured at the 10H beam hole on HIFAR with a 2 inch square collimator in place. These results are in reasonable agreement with calculated values, but the ratio of thermal neutron flux and gamma dose rate is discrepant by a factor of two. Calculations show that an adequate epithermal beam could be obtained if a liquid argon filter is used[31].

Measurements have been made with a one metre liquid argon filter in a dummy fuel element top plate experiment at HIFAR. Various other filters were used as well. Good agreement in flux shape was obtained between calculation and experiment for six detector configurations sensitive to different parts of the neutron spectrum[48]. However, calculations significantly overestimate the neutron flux. In situ experiments are required in 10H to establish epithermal parameters more directly. The self screening foil technique is being used to evaluate the epithermal flux below 1 keV.

HEAVY WATER THERMAL NEUTRON CAPTURE THERAPY

Clinical and patient interest in NCT has reached new heights in Australia following the Fourth International Symposium on Neutron Capture Therapy for Cancer in December 1990 and the publication of review papers in several Australian journals[33,34,35]. This year four Australian patients have gone to Japan for the treatment of high grade brain tumours. These patients take about one litre per day of heavy water ad libitum per oral over several days and receive 4 litres iv at the reactor face. FTIR measurements of urine and blood samples show that up to 23% light water replacement by heavy water can be achieved[36]. Consequently, enhanced thermal neutron penetration is obtained which can add 1 cm to the advantage depth for a wide aperture field.

The effects of heavy water on the advantage depth (depth at which the tumour dose equals the maximum tissue dose) for an 8.65 cm aperture, 30 ppm boron in the tumour and 3 ppm boron in normal tissue, are given below for thermal and epithermal beams[36].

Table 1

$[D_2O]$ %	beam	advantage depth (cm)
0	thermal	4.8
25	thermal	5.8
0	epithermal	8.6
25	epithermal	10.3

A calculational model is needed whereby the required heavy water and fluence conditions can be defined for a given tumour location and blood boron concentration. Thus the potential of thermal NCT to satisfy the neutronics conditions can be determined and appropriate advice given to interested patients.

Heavy water would also be of value in epithermal NCT as it would significantly reduce the dip in thermal flux and therapeutic ratio at the centre of the brain.

AUSTRALIA-JAPAN WORKSHOPS ON MELANOMA

An important strategy in the development of the NCT program was the promotion of joint workshops under the Australia-Japan Bilateral Agreement on Science and Technology. These workshops served to stimulate intranational as well as international research collaborations. Four workshops were held in this series, with joint recommendations to both governments and full publication of papers in the second workshop. Agreements were reached regarding the joint participation in clinical trials for advanced, recurrent melanoma[37] but the selection of suitable patients proved to be a problem.

NCT ADVISORY COMMITTEE

An Ansto advisory committee of scientific and medical specialists has recommended that an epithermal NCT facility should be installed at HIFAR, and that it would be used extensively for the treatment of high grade glioma and melanoma metastatic to the brain if an adequate beam were available. On this basis engineering estimates have been obtained which cost the facility at about two million dollars (AUD). Treatment cost is estimated to be about $10000 AUD for a schedule of 5 dose fractions. Such a facility would cater for the expected demand in Australasia on a single shift basis.

Australian patients may continue to seek overseas treatment in the interim. This committee therefore seeks to determine whether patients are suitable for NCT by possibly testing that the glioma does take up BSH in presurgical administration, and by calculating that neutronics requirements will be satisfied for the depth and size of the tumour. Follow up of the patient's progress post-therapy is also required.

OUTSTANDING QUESTIONS IN BNCT

There are a number of important questions in BNCT that need to be addressed by further research. A brief discussion follows.

Killing of Clonogenic Cells

The average clonogenic cell needs about 10 boron-10 reactions to cover 99% of cell kills, but statistical probability is such that some cells will not receive a lethal dose.

If 99% necrotic cell kill occurs in a large tumour, then the surviving cells must exist amongst the debris of the dead cells. Under these conditions toxic compounds may be released from cell degeneration, capillaries may be truncated and extra vascular transport of nutrients may be inhibited. This scenario would appear to be consistent with the observed success of in vivo NCT where there is no evidence for uniform boron-10 distributions[24].

However, for isolated groups of cells or for lesions where apoptosis (i.e. programmed cell death) is the major cause of cell death, then all target cells must be killed by the boron reaction.

This question can be further investigated by a direct comparison of post NCT in vivo/in vitro colonies and in vivo recurrence rates.

Boron Incorporation in Cancer Cells

The maximum dose to normal tissue can be estimated well enough by serum boron concentrations and relationships to normal tissues, but there can be no certain relationship for the boron concentration in the tumour. In vivo boron magnetic resonance spectroscopy and positron emission spectroscopy can give the average boron content in the tumour, and unless this is adequate therapy should not proceed. However, this decision should be tempered by knowledge of the necrotic or cystic nature of the lesion.

The Effect of Blood Borne Boron on the Microvasculature

This effect has been calculated by several authors and by us but has not yet been subject to quantitative experimental verification. The radiosensitivity of capillaries is the major limitation in radiotherapy, and reduced capillary damage in NCT could be a major advantage of this modality. Injection replication[22] studies pre- and post-NCT are needed to resolve this issue.

Fractionation

The actual gain in low LET repair arises from an effective RBE of 0.5 for gammas for about 5 fractions. However, little quantitative human data is available on low fraction tolerance limits for low LET radiation.

Augmentation of Conventional Therapy

NCT could be used as adjuvant therapy, either pre- or post-radiotherapy. If the therapeutic gain can be significantly increased, then regression may be achieved with minimum risk to overdosing normal tissue. However, a randomised trial would then be needed to determine whether the mixed modality treatment was superior. If any doubts remain regarding normal tissue tolerance in NCT, then adjuvant NCT may be indicated.

Role of Heavy Water in Thermal and Epithermal NCT

It appears that concentrations up to 23% can be achieved without serious effects on the patient[36]. At this level, the AD for an epithermal beam increases from 8.6 cm to 10.3 cm, and in the case of bilateral irradiation, the therapeutic ratio at the midline of the brain is increased by about 20%. Under these conditions, in situ liver might then be treated by epithermal NCT.

SUMMARY

The broad ranging Australian effort in NCT covers virtually all aspects of NCT research. As such, a firm technological foundation has been established which is essential if clinicians are to be convinced of the potential of NCT for the treatment of their patients with poor prognoses. In turn, the clinical interactions provide support for an Australian NCT patient facility. HIFAR offers this potential, but its manifestation requires a more definitive determination of the potential of the 10H beam to deliver a practical therapeutic epithermal beam. There remain many questions in NCT which require further consideration. We hope to address some of these in the near future.

ACKNOWLEDGMENTS

These studies have been supported in part by grants from the Sydney Melanoma Foundation, DITAC, NHMRC, NSW State Cancer Council, Anti-Cancer Council of Victoria, Australian Brain Foundation, Government Employees Assistance to Medical Research Fund, Leo & Jenny Leukaemia and Cancer Foundation.

Collaborating institutions include the Universities of Sydney, Melbourne, Monash and NSW, Peter MacCallum Cancer Institute, Queensland Institute of Medical Research, Royal Melbourne, Royal Prince Alfred, Prince of Wales Children's, Prince of Wales, Prince Henry, Royal Newcastle, Royal Hobart and Westmead Hospitals.

This paper is an overview of the Australian approach to NCT, rather than a global review. We are indebted to the extensive contributions of our international colleagues to this field.

REFERENCES

1. Allen, B.J. Boron neutron capture therapy - a research program for glioblastoma and melanoma. *Aust. Phys. Eng. Sci. Med.* 6, 184-186 (1983).
2. Allen, B.J. In vitro and in vivo studies in boron neutron capture therapy of malignant melanoma. *Proc. First Int. Symp. Neutron Capture Therapy, MIT, Boston, Oct. 1983.* Fairchild, R.G.; Brownell G.L., eds., BNL51730, 341-354.
3. Tamat, S.R.; Moore, D.E.; Patwardhan, A.; Hersey, P. Boronated monoclonal antibody 225.28S for potential use in neutron capture therapy of malignant melanoma. *Pigm. Cell Res.* 2, 278-280 (1989).
4. Wilson, J.G. Synthetic approach to a decarboranyl thiouracil. *Pigm. Cell Res.* 2, 297-303 (1989).
5. Wilson, J.G. Carborane compounds for neutron capture therapy of malignant melanoma. In: *Progress in Neutron Capture Therapy for Cancer.*

Allen, B.J.; Moore, D.E.; Harrington, B.V., eds., Plenum Press, New York, 227-230 (1992).

6. Moore, D.E.; Chandler, A.K.; Corderoy-Buck, S.; Wilson, G.J.; Allen, B.J. Liposomes as carriers of boronated thiouracils for NCT of melanomas. In: *Progress in Neutron Capture Therapy for Cancer*. Allen, B.J.; Moore, D.E.; Harrington, B.V., eds., Plenum Press, New York, 451-454 (1992).

7. Martin, R.F.; Haigh, A.; Monger., C.; Pardee, M.; Whittaker, A.; Kelly, D.P.; Allen, B.J. [157]Gd neutron capture: Potential of [157]Gd-labelled DNA ligands for neutron capture therapy. In: *Progress in Neutron Capture Therapy for Cancer*. Allen, B.J.; Moore, D.E.; Harrington, B.V., eds., Plenum Press, New York, 357-360 (1992).

8. Brown, J.K.; Mountford, M.H.; Allen, B.J.; Mishima, Y.; Ihihashi, M.; Parsons, P. Neutron irradiation of human melanoma cells. *Pigm. Cell Res.* 2, 319-324 (1989).

9. Brown, J.K.; Allen, B.J.; Chapman, J.E.; Mountford, M.H.; Parsons, P. In vitro incorporation of BPA by amelanotic and melanotic murine malignant melanoma cells. In: *Progress in Neutron Capture Therapy for Cancer*. Allen, B.J.; Moore, D.E.; Harrington, B.V., eds., Plenum Press, New York, 369-372 (1992).

10. Ichihashi, M.; Fukuda, H.; Brown, J.K.; Mountford, M.H.; Allen, B.J.; Mishima, Y. In vitro evaluation of BPA for melanoma at Moata in joint work between Japan and Australia BNCT research teams. In: *Progress in Neutron Capture Therapy for Cancer*. Allen, B.J.; Moore, D.E.; Harrington, B.V., eds., Plenum Press, New York, 387-390 (1992).

11. Allen, B.J.; Corderoy-Buck, S.; Moore, D.E.; Mishima, Y.; Ichihashi, M. Local control of murine melanoma xenografts in nude mice by neutron capture therapy. In: *Progress in Neutron Capture Therapy for Cancer*. Allen, B.J.; Moore, D.E.; Harrington, B.V., eds., Plenum Press, New York, 425-428 (1992).

12. Corderoy-Buck, S.; Allen, B.J.; Wilson, J.G.; Brown, J.K.; Mountford, M.; Tjarks, W.; Gabel, D.; Barkla, D.; Chandler, A.; Moore, D.E.; Patwardhan, A. Investigation of boron conjugated thiouracil derivatives for neutron capture therapy of melanoma. In: *Advances in Radiopharmacology*. Maddalena, D.J., ed., 114-127 (1990).

13. Tamat, S.R.; Patwardhan, A.; Moore, D.E.; Kabral, A.; Bradstock, K.; Hersey, P.; Allen, B.J. Boronated monoclonal antibodies for potential neutron capture therapy of malignant melanoma and leukaemia. *Strahlenther. Onkol.* 165, 145-147 (1989).

14. Mallesch, J.L.; Chiaraviglio, D.; Allen, B.J.; Moore, D.E. An intrapancreatic and hepatic model for Boron Neutron Capture Therapy. *AINSE Radiobiology Conference, Sydney, October, 1991*. Abstract.

15. Hill, J.S.; Kahl, S.B.; Kaye, A.H.; Gonzales, M.F.; Vardaxis, N.; Johnson, C.I.; Stylli, S.S.; Nakamura, Y. Tumour localisation of boronated porphyrin in an intracerebral model of glioma. In: *Progress*

in Neutron Capture Therapy for Cancer. Allen, B.J.; Moore, D.E.; Harrington, B.V., eds., Plenum Press, New York, 501-506 (1992).

16. Tamat, S.R.; Moore, D.E.; Allen, B.J. Boron assay in biological tissue by inductively coupled plasma atomic emission spectroscopy. *Anal. Chem.* 59, 2161-2164 (1987).

17. Moore, D.E.; Dawes, A.L.; Allen, B.J.; Bradstock, K.F. Tracking of boron labelled monoclonal antibodies by energy loss spectroscopy in the electron microscopes. In: *Advances in Radiopharmacology.* Maddalena, D.J., ed., 114-127 (1990).

18. Moore, D.E.; Stretch, J.R.; Dawes, A.C.; Cockayne, D.; Allen, B.J.; Constantine, G. Uptake of boronated monoclonal antibodies by melanoma cells visualised by track etch autoradiography and electron energy loss spectroscopy. In: *Progress in Neutron Capture Therapy for Cancer.* Allen, B.J.; Moore, D.E.; Harrington, B.V., eds., Plenum Press, New York, 335-338 (1992).

19. Barkla, D.H.; Allen, B.J.; Brown, J.K.; Mountford, M.H.; Ichihashi, M.; Mishima, Y. Morphological changes in human melanoma cells following irradiation with thermal neutrons. *Pigm. Cell Res.* 2, 345-348 (1989).

20. Barkla, D.H.; Brown, J.K.; Meriaty, H.; Allen, B.J. Ultrastructural changes in tumour cells following high or low LET radiation. In: *Progress in Neutron Capture Therapy for Cancer.* Allen, B.J.; Moore, D.E.; Harrington, B.V., eds., Plenum Press, New York, 395-398 (1992).

21. Elliston, P.; Pope, J.; Allen, B.J. In vivo determination of boron concentrations in rat tumours using boron-11 nuclear magnetic resonance spectroscopy. *EPSM91, Sydney, July 1991.* Abstract.

22. Narayan, K.; Spiropoulos, A. The radio responsiveness of normal and tumour microvasculature. *Australas. Radiol.* 34, 312-316 (1990).

23. Allen, B.J.; Brown, J.K.; Harrington, B.; Izard, B.; Linklater, H.; Maddalena, D.J.; McNeill, J.; McGregor, B.J.; Mountford, M.H.; Snowdon, G.M.; Wilson, D.J.; Wilson, J.G.; Parsons, P.; Moore, D.; Tamat, S.; Hersey, P. Neutron capture therapy research for malignant melanoma. In: *Neutron Capture Therapy.* Hatanaka, H., ed., Nishimura, Niigata, 258-280 (1986).

24. Crotty, K.; Mallesch, J.; Moore, D.E.; Allen, B.J. Histological examination of melanoma xenografts in the nude mouse model; pre and post neutron capture therapy. *AINSE Radiobiology Conference, Sydney, October 1991.* Abstract.

25. Martin, R.J.; D'Cunha, G.; Pardee, M.; Allen, B.J. Induction of double strand breaks following neutron capture therapy in DNA bound [157]Gd. *Int. J. Radiat. Biol.* 54, 205-208 (1988).

26. Martin, R.F.; D'Cunha, G.; Pardee, M.; Allen, B.J. Introduction of DNA double strand breaks by [157]Gd neutron capture. *Pigm. Cell Res.* 2, 330-332 (1989).

27. Allen, B.J.; McGregor, B.J.; Martin, R.F. Neutron capture therapy with gadolinium-157. *Strahlenther. Onkol.* 165, 156-157 (1989).

28. McGregor, B.J.; Allen, B.J. Filtered beam dose distributions for boron neutron capture therapy of brain tumours. *Aust. Phys. Eng. Sci. Med.* 7, 27-34 (1984).

29. Harrington, B.V. Optimisation of an epithermal beam in HIFAR for boron neutron capture therapy. *Ansto/E662* (1987).

30. Harrington, B.V. A calculational study of tangential and radial beams in HIFAR. In: *Neutron Beam, Design, Development and Performance for Neutron Capture Therapy.* Harling, O.K.; Bernard, J.A.; Zamenhof, R.G., eds., Plenum Press, New York, 97-106 (1990).

31. Storr, G.J.; Allen, B.J.; Harrington, B.V.; Davis, L.R.; Elcombe, M.M.; Meriaty, H. Design considerations for the proposed HIFAR thermal and epithermal NCT facility. In: *Progress in Neutron Capture Therapy for Cancer.* Allen, B.J.; Moore, D.E.; Harrington, B.V., eds., Plenum Press, New York, 79-82 (1992).

32. Storr, G. Ideal neutron beam assessment for NCT, submitted.

33. Allen, B.J. Third generation neutron capture therapy for glioblastoma and melanoma brain metastases. *Cancer Forum* 14, 35-36 (1990).

34. Allen, B.J. Epithermal neutron capture therapy: a new modality for the treatment of glioblastoma and melanoma metastatic to the brain. *Med. J. Aust.* 153, 296-298 (1990).

35. Allen, B.J. Potential of neutron capture therapy for the treatment of localised terminal cancer. *Australas. Radiol.* 34, 297-305 (1990).

36. Blagojevic, N.; Allen, B.J.; Storr, G.; Hatanaka, H.; Nakagawa, H. The role of heavy water in dose-depth enhancement in Thermal Neutron Capture Therapy. *AINSE Radiobiology Conference, Sydney, October 1991.* Abstract.

37. Allen, B.J.; Coates, A.S.; McCarthy, W.H.; Mameghan, H.; Mishima, Y.; Ichihashi, M. Thermal neutron capture therapy: The Japanase-Australian clinical trial for malignant melanoma. In: *Clinical Aspects of Neutron Capture Therapy, Brookhaven National Laboratory.* Fairchild, R.G.; Bond, V.P.; Woodhead, A.D., eds., Plenum Press, New York, 69-73 (1989).

38. Constantine, G. Epithermal Neutron Capture Therapy Facility for HIFAR, February 1991; unpublished report.

ROUND TABLE DISCUSSION

APPROACH TO CLINICAL TRIALS OF GLIOMA

Chairperson:
B. Larsson, Zürich, Switzerland

Participants:
B.J. Allen, Sydney, Australia; J.A. Coderre, Upton, New York; R.V. Dorn III, Idaho Falls, Idaho; F. Ellis, Oxford, Great Britain; H. Fankhauser, Lausanne, Switzerland; D Gabel, Bremen, Germany; R. Gahbauer, Columbus, Ohio; D.D. Joel; Upton, New York; Y. Mishima, Kobe, Japan; A.H. Soloway, Columbus, Ohio

LARSSON :

I call this session to order. It is a precious moment. We have limited time and there are many things to be discussed. We should focus our interests on the forthcoming planned neutron capture therapy for supratentorial glioma here at Petten.

The basis for the discussion is well defined. It is the proposal conveyed by Dr. Fankhauser after discussions with his colleagues during the days before this meeting. You all have a copy of this proposal and Dr. Fankhauser made an able presentation yesterday. I think we should dive directly into the subject matter.

However, I would first like to raise one question which I think is very pertinent, especially now, when we have our friends from overseas here. We have promised that we should discuss the strategy and the pace of advancement together, before we are doing anything on human beings. So my first question to the panel would be: Are we right in time, and right in knowledge, for the therapeutic approach that we are planning?

ELLIS :

I think we have to be because we're hoping to start about now. We've got to start with the knowledge and techniques that we have now, and what we have to do is decide how to guide them.

Boron Neutron Capture Therapy, Edited by D. Gable
and R. Moss, Plenum Press, New York, 1992

GAHBAUER :

I think it would still be of extreme importance to continue the normal tissue studies that have been initiated by Dr. Gavin and to continue those on the treatment facility planned in Europe. I believe that these animals have to be studied not only for acute effects, but also for late effects. So, maybe there is a built-in time frame that we may have to wait before clinical trials.

LARSSON :

Of course, the healthy tissue studies that are now initiated and started, would have to be evaluated before the first clinical irradiations, but we have to strike a balance between the wish of avoiding unnecessary time loss and cautiousness.

My question now boils down to whether we have struck this balance or not? Professor Ellis think we have done so, and that's also my personal opinion. But in the case that there are oppositions against this point of view, I would think it is very necessary that we have them exposed today.

JOEL :

I would like to make a comment about compound selection.

We currently have three compounds (BPA, BSH and BSSB), which have been tested extensively in vitro and in vivo, including efficacy of BNCT for controlling experimental intracerebral gliosarcoma.

To my knowledge, BSH is the least effective, yet that is the compound that's being proposed for use in patients at this time. I think the compound issue clearly needs further consideration.

SOLOWAY :

The reason for the great cautiousness in the United States for initiating new clinical trials for the treatment of malignant brain tumour by BNCT is due to the fact that there was clear radiation injury, and in some cases very severe, following BNCT in the last clinical trials 30 years ago. It is therefore absolutely essential that when new trials do begin in the United States one can be assured that there will be both safety and efficacy in the treatment of patients with glioblastoma multiforme. I certainly agree with Dr. Fankhauser that we should not be raising the issue that "We're going to cure brain tumours by BNCT". It is important that this form of therapy should be able to increase useful life expectancy without causing morbidity and mortality.
A second issue which has been insufficiently stressed is that BNCT is really a radiation therapy procedure.

For this reason, I feel that it is mandatory that radiation therapists, especially those who are highly critical of neutron capture therapy, review proposed clinical protocols and offer their criticisms. Criticism, in my judgement, is a very important element in the development of a clinical protocol, especially for a procedure which has had such a tortuous history as BNCT. Neither safety nor efficacy, as Dr. Gabel has stated, cannot and should not be compromised.

ALLEN :

Could I just follow on that point. It seems to me that the gap we have is with the radiotherapy community.

I think this reflects the Australian view. Is the question whether NCT should be considered as an adjuvant therapy, something to be added on to an existing protocol or not?

Now, I'm not sure if that should happen or not, but I think that aspect should be considered as scientifically as we can and if we reject it, then it should be done so on a sound basis. Certainly the risks to NCT would be less, if one uses a dose escalation technique which is complementary to conventional therapy of 30 dose fractions, maybe 20 dose fractions of conventional radiotherapy, followed by a couple of fractions of NCT.

Then slowly increase NCT fractions and use this as a means of a dose escalation technique within basically a framework of existing technology, maybe this has a greater chance of success. I think it would take a lot longer to find out whether the therapy was really effective, but the chances for negative results would be lessened.

CODERRE :

I would like to pursue the compound issue just a moment. I have several thoughts not all necessarily related.

Do not forget that BNCT is being classified as a radiotherapy, but in a sense it is also chemotherapy as well.

It is a binary therapy and I just want to emphasize the burden that is falling upon the chemists working in BNCT to produce compounds. I was jotting down the statistics that Dr. Vermorken displayed that indicated the contribution of chemotherapy to the number of cures. Of only 45% of all cancers that are cured, chemotherapy contributed just 5%, with or without surgery, with or without radiotherapy. BNCT also is chemotherapy. We do not have to deal with the cellular poisons that conventional chemotherapy consists of, but nevertheless, we have to develop drugs that will localize in tumors and there has to be a progression of drugs coming down the pipeline. I think the question before this panel is, are the compounds currently available sufficient for an initiation of clinical trials to establish the principle? But we also always have to have more improved compounds under development and the potential role of industry in Europe and in the United States is to assist us in the compound development and in efficient compound testing. We, in this age of restricted budgets, cannot afford the amount of money that industry puts into developing a single chemotherapy agent, tens of millions of dollars. We have to have an efficient screening process to bring forward the most promising 2nd, 3rd and 4th generation compounds for BNCT.

But, back to the point of Dr. Joel, of the compounds currently available, is BSH sufficient for a demonstration of the principle?

LARSSON :

Was that your point of view that it is sufficient or was it not?

CODERRE :

I think that is a question that needs to be discussed further.

FANKHAUSER :

I have just a comment to this compound issue. As I understand, with BPA, we achieve higher tumour concentrations. But from what I saw yesterday, we have the same level in normal brain and the blood. This was just on one slide, but it was not commented upon. I think this is a very major point because we are now doing a dose escalation study to check for normal tissue tolerance and we use the proposed BSH because we think we are really safe for the normal brain. Could you please comment on this very important point, where, in my eyes, you have the same level in the brain and blood. How could you possibly start with BPA?

JOEL :

Jeff (Coderre) can probably comment more than I, but following BPA administration, ^{10}B concentration in tumor is roughly 4 times that in either blood or normal brain. Thus the efficacy or therapeutic gain, with the other contaminant radiations, will still be in the neighbourhood of 2.5 or so. I agree that with BPA you do have a greater risk of normal brain injury. However, as someone pointed out, it might not be sufficient to treat only bulk tumor. There are islets of cells, protected by the blood brain barrier that BPA might reach, which would not be reached with compounds like BSH. Herein lies a problem. We don't have good microdosimetry or microdistribution techniques, which in my estimation, should be given a high priority.

BSH is going to depend on a breakdown in the blood brain barrier, and the question still remains, as to whether these islets and/or fingers are protected by the blood brain barrier. So there are advantages to each compound. I would agree that if you are going to truly try to not injure normal brain, BSH is a better compound.

I know a lot of work has to be done and if you are going to go ahead at this point, the logical compound is BSH. I felt it needed to be stated that of the compounds that have been tested, BSH is inferior. With BSH, everyone agrees that there is vascular sparing. However, I would be concerned if you are going to irradiate with 50ppm in the blood, because every piece of evidence that I have seen indicates that the critical tissue in the brain is vascular endothelium and not neurons.

DORN :

Encompassing the entire discussion this morning and going back to your original question, I think from a clinical radiation oncology point of view my answer to your question would be that, "Yes, it is the appropriate time to go ahead with clinical trials with BSH and for malignant glioma." Tumor choice has a great deal to do with that and we have had some discussion over the course of the week about what is an appropriate tumor to study as we re-initiate clinical trials in BNCT. I think clearly one has to, in order to simplify things, avoid large numbers of randomized clinical trials. You have to choose a tumor that does not have a good response or cure from any current modality, as opposed to picking a tumour where there is other effective treatment already available. And glioblastoma certainly fits that category. With regard to the compound there is no question that there is experimental discussion and concern about which is the most effective compound. Clearly compound development needs to play a major role over the next several years but none of this research is occurring in a vacuum. I think if you look at BSH, if you look at the dog studies that Pat Gavin has done with respect to safety and response in the dog model, with respect to the studies that Jeff (Coderre) has done at Brookhaven that has shown safety and effectiveness in the small animal model, if you look at the fact that as far as I know no patient of Dr. Hatanaka's has died of BNCT treatment using BSH, I think there is enough evidence to carefully go ahead with clinical trials. Now that assumes that it will be a year or so before the clinical trials get started and the ongoing studies that are being done with BSH in terms of pharmacokinetics in the human and the ongoing radiation safety studies in the animal models, will continue during that year and may affect precisely how those trials are set up. But with those provisos, I think its clearly time to carefully go ahead with BNCT with BSH and glioblastoma.

ELLIS :

I started radiotherapy using radium in 1931. I started, after having seen some cases that had cured cancer in Stockholm, Paris, Germany and England. I had seen them (cases) before, I had seen them during and seen them after, I had however never seen a case right through a course of treatment when I actually started. At that time it was realised that there were difficulties about protection from radium. It was realised that there were dangers to normal tissues through radium and if we had waited to get all these things settled before we started treating patients there would be many patients who died of their cancer who were saved by radium treatment. There is no doubt at all in my opinion, that a lot of the progress has been made in radiotherapy and there has been terrific progress, the cure rates of cancer of the cervix for instance, the cure rates of cancer of the tongue and the larynx, are very much higher than they were in 1931 and if we had waited for all the problems to be settled, we would have never got to the stage where we are now. So I feel that one makes more progress when using things clinically than one does if one just waits for research. The two things can go on together, but with the many types of research that are going on now, if we wait for all these in connection with the meetings that we have had here, we would never get started treating with BNCT. And so, I feel that one has to carry on the clinical work beside that, I don't think we should be here now if it were not for what Professor Hatanaka has done. And that was what first attracted me to BNCT therapy because I heard that he was getting cures of the glioblastoma multiforme and I had never heard of anybody getting cures of that before. In fact, there hasn't been much mention of the fact why he has cures, it has been mentioned by one person, but I mean the question of the blood vessels. He recognised this as a danger and that was the reason for using the BSH, waiting for a time that he determined experimentally in dogs, before using the reactor therapy afterwards. So that some of the problems must have been, I mean they had been settled, possibly not conclusively by Dr. Hatanaka and I think it might be a good idea if he would give permission to allow research workers to go through his records and produce a comprehensive analysis of the work that he has done, at least as support for what we are trying to do now because if he has cures of glioblastoma multiforme, people living long periods of time, I think that in itself is a remarkable thing. It was done like so much in medicine by starting from what one saw, plus theory, plus going on, and a philosophy of waiting and modifying what you are doing as you are going on. Now admittedly I think research is extremely important, and I have fostered it in my own departments as much as possible but it has been in conjunction with the clinical work and that I think is important. If you don't start, you won't get anywhere and it seems to me that if we don't start soon, we may not get the chance to try it, because the project may stop being funded.

LARSSON :

I don't know whether the chairman should express any opinion but I share your point of view in most of what you say. It is my duty to say here that boron neutron capture therapy is in a rather peculiar situation because of its history. A failure now, for example in the form of brain tissue necrosis, may actually not kill only the patient but it may also kill the idea of boron neutron capture therapy.

ELLIS :

I would like to say one other thing, in conjunction with what Dr. Soloway said, I do think we have to have clinical trials and other methods to compare, because it is just not a matter of cure always but of average survival. I think that it is important to choose the right kind of clinical trials because the methods that have been used hitherto are virtually useless in my opinion for doing much for glioblastoma. I think we need to do more to understand the natural history of glioblastoma. Why for instance does a patient, a critical patient who thinks he is cured after a course of radiotherapy, die within six weeks because of rapid recurrence?

There is something different about the way glioblastoma behaves as compared with other kinds of cancer.

So I feel very strongly that clinical trials are different methods for comparison and that will involve very careful and expert statistical advice to make sure of using the cases, that are to be treated, to the best advantage for producing a knowledge of what the results are going to be. I say this, because in 1961, when I was President of the British Institute of Radiology, I started a very simple trial on three days versus five days a week treatment for laryngal cancer and that has only now just been completed. We had 600 cases and it has all been analysed, but the complexity of that very simple trial, has made me very wary of anything that is too complicated.

LARSSON :

Thank you very much! I think you have prepared yourself for this discussion better than anyone of us here.

GAHBAUER:

It is hard to speak after that because we heard a true pioneer speak. I think at some point we have to jump.

I would like to point out a few things. If we look at Hakanaka's work I think what he has done with the thermal beam is akin to an I-125 implant compared to whole brain radiation if we use the epithermal beam.

So I think the potential dangers using epithermal beams treating the whole brain are magnified over what has been done so far.

The other thing is, if we jump into treatment, I think at least we must be prepared that the results will be disappointing. And I think that at that stage we must be ready to have follow-up compounds available to proceed. It would seem logical to me to start animal testing, safety testing in several other promising compounds, to be able to offer a continuous course of clinical evolution of the modality. And another thing that may benefit from such an approach : the toxicity of the compound depends on the total dose. I mean there is nothing to say that it may not turn out to be optimal to use several compounds for the same patient.

We cannot do so unless we have several compounds ready for large animal safety testing.

LARSSON :

I will accept two or three comments on these general issues before we turn over to the details of the protocol.

SOLOWAY :

We need to listen to both our radiation therapists and neurosurgeons as to when clinical trials can commence with malignant brain tumours. We must not be too impatient at the expense of both patients and the entire field. While I can readily appreciate what Dr. Vermorken was saying that there is this pressure to move as expeditiously as possible to begin patient treatment, it is crucial that it not be done prematurely nor that there be unreasonable expectations. It will be necessary for the radiation therapist to examine the clinical protocol and from measurements of the various radiation components to assess the composite dose to tumour, normal brain and especially the cerebral vasculature. Calculations are certainly useful but ultimately these must be translated into actual measurements and total dose. I personally have no emotional commitment to BSH or any other compound. I do think that we are certainly further along in its toxicological evaluation and in its clinical pharmacokinetic studies than with other compounds. However, it should be the responsibility of the radiation therapist, after examining the data and determining the radiation dose which will be delivered to make the decision as to whether clinical trials should commence with BSH or postponed until we possess better targeting compounds. There is confidence on my part that with more chemists involved better compounds will be identified and synthesized. I do feel that there is no longer a question of whether BNCT will work but when it will be used more widely? To achieve this objective we should welcome those radiation therapists who remain sceptical of BNCT and ask for their examination of our data and their criticisms.

LARSSON :

We are approaching the end of this discussion but it is important that the floor also contributes to this essential discussion.

SCHOFIELD :

I would like to support very strongly the line taken by Professor Ellis. I really think we have no alternative now but to go ahead with BSH. What do we say to our funding authorities - "Sorry chaps, we made a mistake, better compounds have come along. Nobody has actually shown that BSH is no good but we think we should wait and try something better."

FANKHAUSER :

My proposal does not include the modalities of BNCT itself, dose and fractionation and things like this, but focused on the aspects around the treatment itself.

LARSSON :

I am fully aware of that, but I also think that in the continued discussion, it would not be very efficient to discuss dose levels, number of fractions, distance between fractions, because these are still open questions.

We have a tentative proposal and the finalization of that proposal would be made in one or one-and-a-half years. So let's avoid these details, although they are very important. They can be used as illustrations of questions and problems, but I think that we should avoid actual figures at this time.

GABEL :

I think I would like to comment on his and Dr. Bartelink's behalf, because we had a very thorough discussion on this and so if I say this, then I hope I represent their views too. In their minds it has not been a question of whether to fractionate or not. They advocate strongly a fractionated treatment. The dose levels of this treatment have to be determined according to the healthy tissue tolerance that we see in dogs and an appropriate backing off of an acceptable damage limit in dogs. Dr. Bartelink mentioned some numbers there.

I think the firm numbers cannot be given now, but we will, in the approach that Drs. Dewit and Bartelink have made, back off from that dose where we see incidents, but we will not back off that much that we actually come into the region where we must expect no efficacy just for satisfying the safety aspect. If you have a very low dose, of course you are safe, but you are definitely not effective. So I think those two points in their view have been clear. I think also in their views, a multi-field, at least a dual-field, irradiation is called for and probably we would not see a single field irradiation. So those are the points that I just wanted to make on their behalf.

MISHIMA :

For my research team, I obtained a body of some 35 people who are nuclear physicists, chemists, pharmacologists, pathologists, radiobiologists, radiotherapists, oncologists and clinicians. We meet twice per year and work together. We are not insisting on who wants to become the first - of course, as a scientist, you have the ambition to be first. But we have to have technical discussions on what is really needed in order to treat the first patients. I think if we could formulate, without destroying our academic independence or freedom, two dozen lines of information from physical dosimetry to chemical dosimetry to phantom dosimetry and so on, I am not sure whether we could say that all glioblastoma multiforme have the same sensitivity or the same repair rate.

As far as melanoma is concerned, melanoma is not a single disease. Five years ago, we thought melanoma was a really simple disease, but we now know that the complexity is in the sensitivity of the melanoma cells to radiation. On the basis of this, we calculate the absorbed dose and determine this with phantom models. We then proceed to treat each case with individual design. Then the body of 10 Department Heads meet to determine what to do with the next case. Secondly, in Japan, there were neurosurgeons who closed one eye when looking at Hatanaka's data. But day by day, we gradually accept what Hatanaka presented. Yet we have a difficult choice when we are facing the patient, depending on the stage of the tumour and depending on its location and depth, and depending on the patient's general conditions, whether this is a suitable case or not. Obviously, from the point of view of the BNCT specialists, we would like to have a first case to treat, which would have even better results. But the medical ethics require that each case should be in an incurable condition. So we are faced with first cases, who have already had surgery and radiotherapy and recurrence and there is no other way to treat them. So this first case is a very tough job. For the second and third cases, if you would have had reasonable success, then society would accept it much easier and you could go ahead with the following case. Thirdly, the effect of BNCT should be objectively observable and presentable with objective data. Then we can convince the medical society as well as the ethical societies, so that the second time will be much easier.

ALLEN :

Could I just make a few other comments before we move on. First of all about BSH. I think all the evidence with BSH is that it is a very safe compound. The capillary effect reduces the effective dose for the most sensitive part of the brain and I think we have heard in this conference that it may be even larger than a factor 3, could be up to a factor of 4. And certainly the uptake by normal brain cells is very low. Now if you compare that with BPA, you have the opposite effect, in one case (BSH) it may be the capillaries, whereas in the case of BPA, it could be the normal tissues. Now with BPA, what happens to the dopamine tracts and the noradrenaline tracts? Whether these centres will take up high amounts of BPA remains to be determined. Whether they will have a high effect after neutron capture therapy and lead to gross impairment remains to be determined. These experiments haven't been done yet. Yet it's highly likely there could be that preferential uptake. I think at this time, whereas BSH has obvious limitations, the chances of success based on the animal work, based on Hatanaka's work, is such that to delay BNCT trials when the current therapy is so poor, is approaching an unethical situation. That is speaking as a scientist and not as a clinician.

There are two phases for the treatment of glioma and hopefully the first phase will show up the problems relating to the second phase. The first phase is in fact the control of the locally disseminated cells close to the tumour. Now the blood brain barrier may be intact in this region. Certainly, elsewhere in the brain, I don't think there is any real doubt that the blood brain barrier is not intact. It takes a lot of cells before the blood brain barrier starts to crack up. But the probable means of transport of the BSH would be extravascular and I expect this is the reason why in fact it works as well as it does. But if local control is achieved in the immediate area of the tumour, then what will show up at a later stage, presumably, is that distant metastases will grow. Maybe the five year life expectancy will not be increased but maybe the one year or two year might be. That would then be a case where you would need to use compounds which were not restricted by the blood brain barrier and BPA is certainly one of those. But again, I think that at this stage, BSH would be safer. We know more about BSH than we do about BPA. Another point with respect to Dr. Fankhauser's proposed schedule, or perhaps that's the radiotherapy schedule, is the one relating to dose escalation starting at something like 5/12ths the dose relating to some end point in the dog brain. Now it strikes me that that is quite low, I don't really know if in fact the possibilities of getting sufficient dose to the tumour would be reached at that level. I think that this is something that needs to be calculated. To actually start at too low a dose when there is no real expectation of achieving control, I think would be too conservative.

One other thing that worries me again is the proposed 48 hour time interval between fractions. Glioma cells can really grow very quickly. I mean, as has been stated, it is incredible what can happen in a matter of weeks after a radiotherapy treatment. If the doubling time for these cells was something like 6 hours then you are throwing away one logarithm of regrowth. It may be longer, maybe 12 hours or 24 hours but it could be factors of 2 or 4 or even factors of up to 10 regrowth of the tumour population. I think that that is something which you can't afford to give away. It is an advantage which we are losing. It is already stated that NCT offers an enormous advantage because the treatment time is coming down from 6 weeks to maybe 1 week. And that in itself could be a major factor in controlling the growth of the secondaries. But I would still like to emphasize the need to minimise that period. The time required for the repair from low LET radiation is something like 6 hours. That really represents the minimum useful time interval between fractions. Now there are other considerations, such as the concentration of the BSH in the blood and tumour-to-blood and tumour-to-tissue ratios, but I think they really have to be balanced out very carefully and not throw away one of the major advantages of NCT. Again, this may be a reason why the Hatanaka single dose work has been effective in the more superficial tumours. I think we have to understand that we don't really know why Hatanaka has the outstanding success that he has achieved in the superficial glioblastomas, where 7 out of 12 of his patients in that group have lived for more than 5 years. I don't fully understand why that has been so successful. But there are a number of factors involved and there is more than just the boron concentrations. So I think these things should be considered. But I would also agree with clinical people here that we know enough to go on, and if we do not go on, I think, that would be incorrect.

LARSSON :

I would like to end the general discussion and focus on the proposed protocol worked out by our group of clinicians. At the present time we should continue on our course and we shouldn't change it but be prepared to accept modifications when new observations come along. I am sure that we now already have a wide variation around the "average opinion". Nevertheless, I interpret the situation as if most of us were agreeing on this course. I am bold enough to make that statement and I would like to read the first comment of the protocol which is so essential: "The main aim of this Phase I study is to determine normal tissue tolerance to the Petten epithermal beam in patients with supratentorial malignant gliomas at known blood levels of ^{10}B. In addition, initial information would be gained on efficacy of BNCT against regrowth of maximally resected tumours and growth continuation or regression of partially resected or biopsied tumours." I think it is very essential that as a group, we agree on these goals. Because if you start to consider what we could do, given all the time, given all the economical resources, we would probably choose a more sophisticated approach. I would like to first have your points of view concerning the general aims of the Phase I clinical trial here at Petten.

I think the arguments were laid down very clearly by Dr. Fankhauser yesterday. He stressed the point of view that we should not be disappointed if the proposed combination of BSH and the epithermal neutrons at Petten would not lead to the expected results. Further sophistication will come in when we are going to discuss Phase II and Phase III and comparative studies with other modalities, with or without adjuvant therapy. So I take it that you accept the goals of the study. I have heard comments in the intermission, that we should concentrate more on the scientific aspects of this material. There are many things to be measured, many things to be studied in these patients. So what should be pursued are the supplementary studies with more or less sophisticated techniques. I mentioned one example yesterday, i.e. the PET camera studies. It came to my mind that one parameter that permits grading of the brain tumours is the receptor density for somatostatin and EGF. I think that we should actually have respect for the contributions of molecular and cellular biology and try at least qualitatively to introduce as early as possible, the biological basic aspects into this clinical study. This would lend greater respect to the study among critical radiotherapists all over the world. It would also permit scientific contributions to join the goals of this study. So I am just trying to interpret what I heard in the intermission.

SAUERWEIN :

You (Fankhauser) propose for group 1, a re-operation in case of occurrence of radionecrosis. You don't propose it for groups 2 and 3. For group 3 it is clear, but for group 2, I think we should also discuss whether, if radionecrosis occurs, these patients should also be operated. I think it is very important that we have the maximum amount of information on what has happened in the brain. So we should really write into the protocol that a re-operation, if radionecrosis occurs, should be the first therapeutic modality to be discussed.

FANKHAUSER :

The reason that re-operation has not been proposed in the protocol for group 2 is because these patients have been operated once already and we could not remove the tumour totally. It is therefore unlikely that you will propose a second operation. If we introduce the operation for group 2 patients, then this group will be equal to group 1. It is not forbidden to re-operate on these patients and, as someone pointed out, even group 3 patients are sometimes re-operated. For instance, we could make a stereotactic biopsy if we think that it is not reasonable to operate this patient. We try radiotherapy, chemotherapy and then when we see that this treatment is a failure, we may say "Now we have nothing to lose" and we operate. Therefore you are allowed to operate on these patients but it is not part of the protocol. In any case we will have some information concerning recurrence and radionecrosis from autopsy.

ELLIS :

I don't think there is any real distinction between groups 1 and 2 judging from results in the past. Group 1 is really a partial resection because they always recur and I wonder if it is worthwhile making that distinction.

Again aiming at simplicity in any trials.

DORN :

I would add my support to Sauerwein as well. I think as I mentioned the other day, if you would include it in the protocol under group 2, even though you didn't feel that you could go for total removal, one could put in the protocol that if there is recurrence of radionecrosis then one should consider at that point, whether or not surgery is a viable option. This would not be mandated, but just suggests that that should be considered.

FANKHAUSER :
Yes, I think that I mentioned that re-operation is allowed but it is not a requirement. But to answer Prof. Ellis, we distinguish between group 1 and group 2 because in group 1 the re-operation is compulsory and I think it is useful if you can inform the family right from the beginning what will happen. If you start BNCT and after BNCT you tell them it has failed and we have to re-operate, they will be very disappointed.
So it is for the sake of informing early those patients who in all cases will be re-operated.

ELLIS :
It would follow from what I said that you would re-operate on all of them anyway, if it was indicated clinically. There isn't really any distinction fundamentally between groups 1 and 2.

SAUERWEIN :
I absolutely agree with Dr. Dorn. The next point I would like to ask to Heinz (Fankhauser) or suggest to the panel is the following: We are able now to follow boron by MRI spectroscopy, also in humans. Could we do this in our patients? If we repeat the BSH application before each fraction, we could get more information about what is happening with the blood brain barrier. We could try to have a MRI spectroscopy in these patients, after the third irradiation for example.

ALLEN :
I would like to follow up on that point. I guess one of the problems about this protocol is the delayed long term effects, which will take 6 to 12 months perhaps to show up, and the ability to identify radionecrosis or edema, separating edema or other factors. Morphological changes take a long time to occur. Positron emission tomography allows you to identify functional changes very quickly and although not every one can go through a PET facility, I think it should be a requirement that at least one stream of patients have passed through positron emission tomography, so that radionecrosis can be identified at the earliest possible time to allow for modification of the trial. Waiting 6 or 12 months is, I think, really far too long and certainly to my knowledge positron emission tomography can make that distinction very quickly and has been used in radiation therapy to that effect.

FANKHAUSER :
What will the PET scan show when you have at the same time tumour recurrence and radionecrosis which can definitely occur?

ALLEN :

It is a long time since I looked at this but I believe it would show a difference. You can identify the difference.

LARSSON :

I can corroborate this. After the PET meeting in Uppsala two weeks ago,
this was a major issue. So I think that it is quite possible, especially if we include receptor studies. It is by no means very difficult. But we still haven't
seen, for example EGF labelled with bromine-77. There may be some technical difficulties but as a preparation for phase II and phase III studies,
preparations for such studies should be made now.

DORN :

I would like to make one suggestion. About halfway down (the protocol), in
all three groups concerning the pattern of care for glioblastoma, I would suggest that the maximum delay that you have built in to all three groups
should be decreased. I would decrease that to an absolute maximum of four
weeks at the most, and less if you could. In the current course of care of a
glioblastoma patient, in my own practice, once you make a diagnosis of
glioblastoma, and recognizing the rapidity with which it can grow, you tend
to move right on to treatment as soon as possible. Now I know there will be
problems with the logistics of BNCT but if we allow a delay of 6 weeks, we
may be setting ourselves up for a situation (even though this isn't an efficacy
study), where the patients aren't going to do as well because there is a longer
delay to treatment than they would normally see in standard care.

FANKHAUSER :

I absolutely agree, and I myself proposed 4 weeks but it was felt that it
might be difficult. I can tell you that even with conventional radiotherapy, we
now have difficulties in our Radiation Department in Lausanne. We had some
patients who had to wait for more than 6 weeks and some got growth recur-
rences before they started radiation therapy. So I absolutely agree that we
have to reduce this delay, but I don't know how we can achieve this.

ELLIS :

Does this mean that anyone suitable for the trial who, through no fault of
the people concerned, was not available for treatment in 4 weeks would be
excluded. I think that could possibly cut down a good number of patients. We
can say "as soon as possible" and put a maximum of 6 weeks. But I think that
the logistics, arrangements, etc., especially if holidays intervene or if there
are strikes and things like that, we have to be careful how patients are accepted and then excluded, because an exclusion could also mean losing
someone completely who is waiting for treatment and might thus be excluded
for more than 6 weeks. I feel one has to be careful about that.

LARSSON :

I am grateful for this comment, the implications are obvious. In that context we should go back to the question raised by Dr. Allen about the fractionation pattern: the 48 hours versus 24 hours issue. Are there comments on these points?

FANKHAUSER :

This issue of the 6 weeks to me seems more important than the issue of the 48 hours.

GABEL :

The radiobiology would I think, indicate that at least 6 hours are to pass between fractions, if I have read my books correctly. The problem, if we wait for longer periods, has been clearly demonstrated in the work of Heinz Fankhauser and Dietrich Haritz. There is a reduction in blood contents of boron, but also in the tumor contents of boron. They may go in parallel after say about 24 hours, but definitely the absolute amounts of boron will decrease over time and that means that if you wait for a long period you might have to re-administer boron. When we do so, we know that we have to wait another 24 hours to achieve a good ratio again between whatever presumable tumor might be left, that we don't see, microscopic or macroscopic tumor, and the blood and that gives us the minimum period of 24 hours between administration and irradiation. Of course we could think, that we wait those minimum 6 hours plus perhaps 2 or 3 hours or so and repeat the second fraction without re-administering boron but that must then be done without re-administering boron. Then of course, this could be a viable alternative. We would have to look more carefully into the pharmacokinetics after 24 hours to actually say whether this is viable or not. But I think we are not locked to those 48 hours. Those 48 hours would be required if we want to re-administer and I think that is sure.

LARSSON :

With BSH, maybe 48 hours may be motivated, but those compounds that we are looking forward to may permit longer retention times. In that case we would perhaps be happy to have a material to lean on in Phase I which was based on 24 hours rather than 48 hours. I think also that one has to look at the economical aspect. If patients could be treated in one week rather than in two weeks, I think it is highly important. It has also human implications because patients coming here from far away, would certainly like to go back to their families as soon as possible. So I think one should think once or twice before one actually accepts doubling the total treatment time.

ELLIS :

I am not quite sure what we are talking about. Are we talking about intervals between fractions or intervals between BNCT and the radiation.

LARSSON :
Intervals between fractions.

ELLIS :
I think both ought to be fairly rigid. Because otherwise there is going to be an element of doubt entering into considering results. But it should be possible to book for instance, the treatment dates on the reactor and having booked that, then the time for giving the BNCT should be adjustable. I mean the interval shouldn't be adjustable, but the time should be easy to fix. The important time would be the reactor time.

Also, we ought to decide on what is the optimum time between fractions, taking into account many things.

Much information is available. It is not a thing to discuss here except to say that it ought to be settled and kept as fixed as possible so as to avoid complications in analysing the results afterwards.

LARSSON :
I think this question cannot be settled here. One thing I don't know, is whether there is a risk for increased intracranial pressure with the shorter time and the rather large field of irradiation?

ELLIS :
It has been shown for instance, that the birth rate of glioblastoma cells can be as high as 1.7 per 100 cells per hour and that means that at the end of the 6 week course for instance, instead of having about 1 cell or no cells as you should have, you have got 3 million.

I mean that's the kind of thing one is faced with and I think the question of the interval between fractions has to be carefully considered in detail by the ultimate planners of the protocol.

LARSSON :
Very good. I think that is an important message.

WATKINS :
I have a point about this delay between fractions, there are a lot of contentious issues here. I've heard various arguments and some I agree with and some I don't, but one thing that hasn't really been mentioned, that I know that Pat Gavin and some of the other people have put forward and shouldn't be forgotten, is that the vascularity of these tumours varies between one day and the next. If you are going to give BNCT on one particular day then maybe you have some particular vascular system within the tumour open. If you then give using that same administration of compound, a second fraction of radiation you still have the same vascular system open. Surely it is better to wait some time, a day or so, and give a second dose of compound. I think that is one particular aspect that hasn't really been discussed at all. Maybe

I'm completely wrong, I'm talking way out of my field here, but is this an aspect that should be considered?

ALLEN :
   Could I just make a comment to that. There are basically two grades of targets here. There is one where the tumour has been completely excised and what you are looking at are those peripheral cells in which, I would think the vasculature doesn't play any real role. Boron transport would be mainly extravascular. And in the second and third group of tumours in the protocol, the vascular system is intact. What you are saying may be quite true. So in principle there could be a different protocol to be applied because we are actually looking at different targets.

GABEL :
   These are two questions that Peter Watkins addressed. The question of fractionation in terms of conventional radiotherapy involves the four R's of radiotherapy. That is, repair, re-oxygenation, re-distribution and re-population, and I think those would favor any schedule of fractionation. There is a fifth R in boron neutron capture therapy which is re-targeting and that can only be achieved if you re-administer your compound. If it has passed by and it is retained by those cells that you want to get but not by all the cells, the only chance to get at those cells again is re-targeting. So I think this of course must be taken into account. So re-targeting would argue for renewed administration of the boron compound whereas the other four R's are more or less taken care of by any kind of reasonable fractionation. There is one other point of course that we need to take into account when we talk about one fraction every two days, or two fractions on one day and that is the logistics and the capacity of a facility like this. Because if you say you do not want to have more than 8 or 10 hours between fractions, that actually would mean that you have to work two shifts on that day if you want to do this treatment properly. I mean, if you assume an 8 hour working day and everybody is finished after 8 hours, a new team has to come in after 8 hours to take over and that is certainly something that we need to take into account. What then could be the capacity of this facility and the capacity of the people dealing with this facility.

ELLIS :
   I feel that we have to consider what we are doing. We are using an alpha particle to kill the cells. Now with the high LET radiation, it doesn't seem to matter whether you fractionate or not. In that case it would be easier and quite logical it seems to me, but I am willing to be convinced by the radiobiologists, that we should concentrate on one treatment and not on fractionated treatment. But I hope I am not throwing a spanner in the works.

LARSSON :

There have been many comments of this type in the past and I think these few fractions represent a reasonable compromise.

MISHIMA :

I only treated one case with a 2-fraction treatment, a melanoma occurring at the tip of the third finger, because I was afraid to give sufficient initial dose, as necrosis of the fingers may have developed. So we gave about 75% of the dose. Then we did many animal experiments. We reconfirmed that we have to be very careful for the second or third administration of ^{10}B-BPA or BSH. The uptake is not the same. Usually we can cure the lesion because of high LET particle irradiation and there is no repair. We have ended with single irradiation to eradicate the tumour.

LARSSON :

Starting with Dr. Dorn, I will now end this session by letting everyone in the panel giving one pregnant advice to this European concerted action. Maximum 30 seconds.

DORN :

My comment would pick up where Dr. Ellis and Prof. Mishima left off. A truly minority opinion, but I believe that, and I understand if you don't want to do this right in the protocol, the European group or one of the US groups very quickly needs to do the experiment to find out whether BNCT is better in single doses or fractionated because of the re-targeting "R" that Detlef (Gabel) mentioned. I don't think we really know which of the two is best and I think that we have to determine that.

CODERRE :

I would like to address my 30 second comment to an extension of what Prof. Mishima was saying. We do not know, with any compound, what the uptake in the tumor is after one fraction or two or three. The question is whether, in these wounded tumors, you have affected the vasculature or not? We have some indirect evidence that perhaps these sulfhydrylboranes are killing cells through vascular damage. What is that going to do to the tumor uptake? It could increase, it could decrease it, and I think your dosimetry for each fraction is based on the assumption that the uptake is constant.

ALLEN :

That may be true but it is a little irrelevant. What is relevant is what is happening to the normal tissues. We know that the gamma component to the normal tissues will be halved effectively by 4, 5 or 6 fractions. It is my own feeling, although I would certainly agree that very little experimental work has been done, that variation in tumour biology and in cell-line properties is enormous. I would have much more confidence in the fractionated approach from the point of view of getting every cell or getting most cells, than I would be in just one treatment. Statistically and morphologically, could you expect to get all your boron into the cancer cells when you have large areas of necrosis? I think it is not a reasonable proposition. In some types of tumours it is perfectly reasonable. But I think in most tumours the fractionated approach would be best.

MISHIMA :

You are discussing only BSH, but with this I do not agree. Prior to irradiation planning and using BSH, you have also to think about new compounds and to invite a neurochemist. The metabolic pathways are very unique to the glioblastoma cell. We are talking about the uptake of BSH, but the retention time of BSH is very important. For malignant melanoma we have a retention time which is very high, because this compound is taken up inside of the polymer of each melanosome. So you can have multiple injections and easily accumulate tumour to blood ratios of more than 10. So prior to this, I think you should invite many fresh people to acquire knowledge about the neuron and our glioblastoma cells, try to make efforts to synthesize newer compounds. Secondly, of course, the excitement for monoclonal antibodies died down some 2 or 3 years ago, but now a revival in this is coming up, through knowledge of the antigen structures. So by using gene technology, you can determine the antigenic structure sequence. When you synthesise a peptide, where you have a high affinity of antibodies, you can create highly immunogenic antibodies recognised not by the mass but by the tumour. We must try not to give up hope. But I think we have to make efforts for new horizons also.

SOLOWAY :

I should like to make just one piece of advice with my clinical colleagues before we embark upon BNCT protocol in patients. I think you may wish to present your rationale, data in large animals and the clinical protocol to respected radiation therapists who are not engaged in BNCT. Their review and suggestions may be of great benefit in avoiding pitfalls that are possible in any experimental radiotherapeutic procedure.

GABEL :

I don't want to comment on our own proposal. I would rather like to draw everybody's attention to the fact that the European Collaboration for Boron Neutron Capture Therapy has not only the goal to initiate clinical trials with glioma but also to develop new compounds for this and maybe every one of us, every now and again, forgets that we are not a single goal association. I think we do need more effort in all the other aspects as well.

FANKHAUSER :

At the beginning, Barry Allen made a suggestion. He was scared to give a new treatment and an unknown dose to the initial patients. So he suggested that we should add BNCT to conventional radiotherapy and slowly increase it. This is an important question. I am really wondering why nobody has supported this. It looks as if people today agree that we shouldn't mix treatments. But I am afraid that once people are back home, this idea may come back. I would like to advise very strongly against these two modalities. It is true we will give an unknown dose to the patients, but giving a known dose plus an unknown dose is just the same as giving an unknown dose alone. If we mix the two modalities from the beginning, we will prevent straightforward analysis.

JOEL:

Since I won't win the compound argument, I would like to express my belief that a top priority should be the development of techniques for the micro-localization of these compounds. This will give us critical information about any compound that we simply don't have right now.

ELLIS:

The essential argument is whether a lethal tumour dose is possible to the whole brain without destroying normal tissue or almost as important, without destroying intellectual function. I think that needs considering, but we know that ordinary gamma radiation or X-rays hasn't produced any results. With BNCT, we are bound to give some gamma radiation. But we are not treating with an unknown dose if the tolerance dose has been determined in animals and determined by comparison with doses that have been given to people in the past over many years. There is a lot of experience with radiation given to brains, for cranial meningioma, medulloblastoma and other cerebral malignant tumours, so that I think things like that should be possible to decide without further experiments in view of the fact that we have a reasonable RBE average.

The question of the boron compound transport, one thing that I remember was mentioned to me by Professor Bleehan of Cambridge, is that radiosensitisers within a short time, got the same concentration in the tumour as in the blood. I wondered if that had been considered as a possible method of transporting boron into tumours. That's a technical point which I just get in because I remember it, and I think we should consider very carefully the natural history of the glioma cells. Why do they divide so rapidly making the gap between fractions and waiting time so important?

GAHBAUER:

I believe that it is very important to look carefully at the normal tissue damage in these large animal studies.

I feel that we should start doing this also on any new promising compound to allow for continuity of clinical research should the results with the BSH be disappointing. I agree with Dr. Fankhauser that we should use BNCT without conventional radiation therapy. I believe fractionation is important for the reasons explained.

The Low LET component is significant, therefore fractionation allows the delivery of a larger High LET dose. Of course, re-targeting may provide another rationale for fractionation. Since the main therapeutic dose to the tumour is delivered by High LET radiation, which is relatively insensitive to fractionation effects, some dose escalation can be accomplished in BNCT by starting with a large number of fractions. When one uses 10 fractions instead of 4, one would basically de-escalate the effective normal tissue dose somewhere between 30 and 40% without any real detriment to the tumour effect.

LARSSON :
Thank you very much! I hope you found it worthwhile. I thank the panel, I thank the audience.

AUTHOR INDEX

PARTICIPANTS

Ahlf, J.
CEC-JRC-IAM
Postbus 2
NL-1755 ZG Petten
Netherlands

Ait Abderrahim, H.
SCK MOL
Boeretang 200
B-2400 Mol
Belgium

Alberts, R.
Univ. of Bremen
Dept. of Chemistry
P.O. Box 330440
D-2800 Bremen 33
Germany

Alberts, U.
Albersstrasse 12
D-2800 Bremen 1
Germany

Allen, B.J.
Lucas Heights Labs.
Appl. Physics Div. AAEC
PMB 1
AUS-2234 Menai NSW
Australia

Alpen, E.L.
Univ. of California
Lawrence Berkeley Lab.
MS-10/110
Berkeley, CA 94720
USA

Angel, R.M.
Centronic Ltd.
Centronic House
King Henry's Drive
New Addington
Croydon CR9 0BG
U.K.

Auterinen, I.
Techn. Research Centre Finland
Reactor Laboratory
P.O.Box 200
SF-02151 Espoo
Finland

Bartelink, H.
Ned. Kanker Instituut
Plesmanlaan 121
NL-1066 CX Amsterdam
Netherlands

Begg, A.
Ned. Kanker Instituut
Plesmanlaan 121
NL-1066 CX Amsterdam
Netherlands

Blattmann, H.
P.S.I. Villigen
CH-5232 Villigen
Switzerland

Burian, J.
Nuclear Research Institute
CS-25068 Rez/Prague
Czechoslovakia

Carlsson, J.
University of Uppsala
Dept. of Radiation Science
P.O. Box 535
S-75 121 Uppsala
Sweden

Casado, J.
CEC-JRC-IAM
Postbus 2
NL-1755 ZG Petten
Netherlands

Ceberg, C.
Lund University
Inst. for Radiofysik
Lasarettet
S-22185 Lund
Sweden

Chiaraviglio, D.
Via A. Scarpa, 11
I-27100 Pavia
Italy

Coderre, J.
Brookhaven Natl. Lab.
Medical Department
30, Bell Ave., Bldg. 490
Upton, NY 11973
USA

Constantine, G.
JRC Petten
P.O. Box 2
NL-1755 ZG Petten
Netherlands

Crawford, J.F.
P.S.I. Villigen
CH-5232 Villigen
Switzerland

de Haas, J.B.
ECN
Postbus 1
NL-1755 ZG Petten
Netherlands

De Raedt, C.M.J.
SCK MOL
Boeretang 200
B-2400 Mol
Belgium

Dorn III, R.
EG&G Idaho Inc.
Idaho Nat. Labs.
P.O. Box 1625
Idaho Falls, ID 83402
USA

Ellis, F.
2 Bladon Close
Woodstock Road
Oxford OX2 8AD
U.K.

Ensing, G.E.
Mallingkrodt Diagnostica
Westerduinweg 3
NL-1755 ZG Petten
Netherlands

Fankhauser, H.
Univ. de Lausanne
Centr. Hospit. Univ. Vaudois
Dept. of Neurosurgery
CH-1011 Lausanne
Switzerland

Farnworth, C.R.
Centronic Ltd.
Centronic House
King Henry's Drive
New Addington
Croydon CR9 0BG
U.K.

Freudenreich, W.
ECN
Postbus 1
NL-1755 ZG Petten
Netherlands

Gabel, D.
Univ. of Bremen
Dept. of Chemistry
Postfach 330440
D-2800 Bremen 33
Germany

Gahbauer, R.
Ohio State University
Div. Neurologic Surgery
410 West 10th Ave., N-911
Columbus, OH 43210
USA

Gavin, P.
Washington State University
College of Vet. Medicine
McCoy Hall
Pullman, WA 99164
USA

Gibson, J.A.B.
Atomic Energy Res. Establ.
Bldg. 775
Harwell
Didcot, OX11 ORA
U.K.

Griebenow, M.
EG&G Idaho Inc.
Idaho Nat. Labs.
P.O.Box 1625
Idaho Falls, ID 83402
USA

Harfst, S.
Univ. of Bremen
Dept. of Chemistry
Postfach 330440
D-2800 Bremen 33
Germany

Haritz, D.
ZKH St. Jürgen-Straße
Neurochirurgie
St. Jürgen-Straße
D-2800 Bremen 1
Germany

Harrington, B.
Australian Nucl. Sc. & Techn. Org.
Lucas Heights Research Lab.
New Illawarra Road
Lucas Heights
Australia

Haselsberger, K.
University of Graz Med. School
Dep. of Neurosurgery
Auenbruggerplatz 5
A-8036 Graz
Austria

Hawthorne, M.F.
UCLA
6115 Young Hall
405 Hilgard Avenue
Los Angeles, CA 90024
USA

Hemler, R.
Kliniek voor Kleine huisdieren
Geneesk. Gezelschap Dieren
Postbus 80154
NL-3588 TD Utrecht
Netherlands

Herrmann, H.D.
Neurologische Univ. Klinik
Abt. Neurochirurgie & Polikl.
Martinistrasse 52
D-2000 Hamburg 20
Germany

Hoffmann, U.
Univ. of Bremen
Dept. of Chemistry
Postfach 330440
D-2800 Bremen 33
Germany

Huiskamp, R.
ECN
Postbus 1
NL-1755 ZG Petten
Netherlands

Joel, D.D.
Brookhaven Natl. Lab.
Medical Department
30, Bell Ave., Bldg. 490
Upton, NY 11973
USA

Kahl, S.B.
University California
Dept. Pharmaceut./Chemistry
S-926
San Francisco, CA 94143
USA

Kalef-Ezra, J.
University of Ioannina
Medical Physics Laboratory
Medical School
GR-451 10 Ioannina
Greece

Ketz, H.
Univ. of Bremen
Dept. of Chemistry
Postfach 330440
D-2800 Bremen 33
Germany

Konijnenberg, M.
Ned. Kanker Instituut
Plesmanlaan 121
NL-1066 CX Amsterdam
Netherlands

Krüger, U.
Schering AG
Dept. Contrast Media Chem.
Muellerstrasse 170-178
Postfach 650311
D-1000 Berlin 65
Germany

Larsson, B.
University of Zurich
Paul-Scherrer-Institute
Inst. f. Medical Radiobiology
August Forel-Strasse 7
CH-8029 Zürich
Switzerland

Loughlin, T.
Wessex Neurological Centre
Southampton General Hospital
Tremona Road
Southampton
U.K.

Mijnheer, B.
Ned. Kanker Instituut
Plesmanlaan 121
NL-1066 CX Amsterdam
Netherlands

Mishima, Y.
Mishima Inst. of Dermatological
Research
17-8-801, 3-chome
Motomachi-dori, Chuo-ku
Kobe 650
Japan

Moore, D.
University Sydney
Dept. of Pharmacy
Sydney 2006
Australia

Morris, G.M.
Churchill Hospital
Research Institute
Dept.Radiother. & Oncol.
Old Road
Oxford, OX3 7LJ
U.K.

Moss, R.
CEC-JRC-IAM
Postbus 2
NL-1755 ZG Petten
Netherlands

Orenstein, D.
Universite Louis Pasteur
Service de Neurochirurgie
CHU Hautepierre
F-67200 Strasbourg Cedex
France

Papaspyrou, M.
Institute of Medicine
Research Centre
Postfach 1913
D-5170 Jülich
Germany

Pellicer, J.A.
University of Valencia
Dept. de Biologia Animal
E-46100 Burjassot (Valencia)
Spain

Perks, C.A.
AEA Technology
Harwell Laboratory
B 364
OX11 ORA Harwell
U.K.

Pettersson, O.A.
Uppsala Univ.
Dept. of Radiation Science
P.O. Box 555
S-75121 Uppsala
Sweden

Pfister, G.
Institut für Kernenergetik
Universität Stuttgart
Pfaffenwaldring 31
D-7000 Stuttgart 80
Germany

Philipp, K.I.H.
ECN
Postbus 1
NL-1755 ZG Petten
Netherlands

Raaijmakers, C.P.J.
Ned. Kanker Instituut
Plesmanlaan 121
NL 1066 CX Amsterdam
Netherlands

Ravensberg, K.
ECN
Postbus 1
NL-1755 ZG Petten
Netherlands

Rave-Fränk, M.
Georg Dehio-Weg 12
D-3400 Göttingen
Germany

Raviv, O.
Univ. of Bremen
Dept. of Chemistry
Postfach 330440
D-2800 Bremen 33
Germany

Ricchena, R.
JRC Ispra
I-27020 Ispra (Varese)
Italy

Rösler, J.
Univ. of Bremen
Dept. of Chemistry
Postfach 330440
D-2800 Bremen 33
Germany

Salford, L.
Lund University
Div. of Exp. Neurooncology
Lasarettet
S-22185 Lund
Sweden

Sauerwein, W.H.G.
Univ. Klinikum Essen
Strahlenklinik
Hufelandstrasse 55
D-4300 Essen 1
Germany

Schofield, P.
Institut Max von Laue-Paul Lan-
gevin
Avenue des Martyrs
B.P. 156X
F-38240 Grenoble
France

Schupbach, D.
Univ. de Lausanne
Centr. Hospit. Univ. Vaudois
Dept. of Neurosurgery
CH-1011 Lausanne
Switzerland

Siefert, A.G.
CEC-JRC-IAM
Postbus 2
NL 1755 ZG Petten
Netherlands

Snijders-Keilholz, A.
Academisch Ziekenhuis Leiden
Gebouw 20
Postbus 9600
NL-2300 RC Leiden
Netherlands

Soloway, A.H.
Ohio State University
College of Pharmacy
500 West 12th Avenue
Colombus, OH 43210
USA

Southworth, G.S.
Boron Biologicals Inc.
2811 O'Berry Street
Raleigh, NC 27609
USA

Stecher-Rasmussen, F.
ECN
Postbus 1
NL-1755 ZG Petten
Netherlands

Stragliotto, G.
Univ. Lausanne
Dept. of Neurosurgery
CH-1011 Lausanne
Switzerland

Sullivan, W.M.
Binary Therapeutics Inc.
44 Wall Street
New York, NY 10005
USA

Sweet, W.
Mass. General Hospital
ACC 312
15 Parkman
Boston
USA

Tovarys, F.
Statni Ustvan Narodniho Zdravi
Roentgenova UL. 1
CS-15119 Prague 5 Motol
Czechoslavakia

Twilegar, R.
Neutron Technology
877 Main Street
Boise, ID 83702
USA

Van Dijk, J.J.
Mallingkrodt Diagnostica
Westerduinweg 3
NL-1755 ZG Petten
Netherlands

Van Doorn, B.
Rijksuniv. Groningen
Fac. of Medicine
Postbus 72
NL-9700 AB Groningen
Netherlands

Van den Aardweg, G.J.M.J.
Dr. Daniel den Hoed Kliniek
Dept. of Radiotherapy
Groene Hilledijk 301
NL-3075 EA Rotterdam
Netherlands

Van der Kogel, A.
Inst. of Radiotherapy
G. Grooteplein zd. 32
NL-2656 Nijmegen
Netherlands

Vecht, C.J.
Dr. Daniel den Hoed Kliniek
Dept. of Radiotherapy
Groene Hilledijk 301
NL-3075 EA Rotterdam
Netherlands

Vermorken, A.J.M.
CEC DG XII
Medical Research Programme
Rue de la Loi 200
B-1049 Brussels
Belgium

Verrijk, R.
Ned. Kanker Instituut
Plesmanlaan 121
NL-1066 CX Amsterdam
Netherlands

Vogt, U.
Univ. of Bremen
Dept. of Chemistry
Postfach 330440
D-2800 Bremen 33
Germany

Voorbraak, W.
ECN
Postbus 1
NL-1755 ZG Petten
Netherlands

Voormolen, J.H.C.
Academisch Ziekenhuis Leiden
Gebouw 20
Postbus 9600
NL-2300 RC Leiden
Netherlands

Warnecke, G.
Klinikum Minden
Neurologische Klinik
Friedrichstrasse 17
D-4950 Minden
Germany

Watkins, P.R.D.
CEC-JRC-IAM
Postbus 2
NL-1755 ZG Petten
Netherlands

Westphal, M.
Neurologische Univ. Klinik
Abt. Neurochirurgie & Polikl.
Martinistrasse 52
D-2000 Hamburg 20
Germany

Wheeler, F.J.
EG&G Idaho Inc.
Idaho Nat. Labs.
P.O.Box 1625
Idaho Falls, ID 83402
USA

Zamenhof, R.G.
New England Med. Center
Dept Radiat. Oncology
Box 246
750 Washington Street
Boston, MA 02111
USA

Zhu, J.
CEC-JRC-IAM
Postbus 2
NL-1755 ZG Petten
Netherlands

Zonta, A.
Policlinico St. Matteo
1st Pat. Chirurgica
Piazzale Golgi
I-27100 Pavia
Italy

SUBJECT INDEX